'An important book, in the sense that it will probably be read by a large number of people'
Journal of the British Interplanetary Society

'A readable and provocative vision of humanity's future in space'
Carl Sagan

'As mind-stretching—and as fascinating—as good science fiction but he is not fantasising. He is genuinely extrapolating from fact'
Evening News

'An interesting book, full of bold speculations presented in plausible shape'
Books and Bookmen

'Heady but exciting'
Daily Telegraph

The Next Ten Thousand Years

A Vision of Man's Future
in the Universe

Adrian Berry

CORONET BOOKS
Hodder and Stoughton

To my wife, Marina

Copyright © Adrian Berry 1974
Introduction © 1974 Patrick Moore
Appendix III © 1974 Iain K. M. Nicolson and Adrian Berry

First published in Great Britain 1974 by
Jonathan Cape Limited

Coronet Edition 1976

─────────────────────────────

Printed and bound in Great Britain for
Coronet Books, Hodder and Stoughton
By Richard Clay (The Chaucer Press) Ltd,
Bungay, Suffolk

ISBN 0 340 19924 5

Contents

Acknowledgments

A large number of people have helped me in writing this book. I am particularly grateful to Dr Anthony Michaelis, who read the manuscript twice and made many useful suggestions. I thank also my parents and my wife for doing the same. Patrick Moore and Iain Nicolson contributed a great deal of their knowledge. I had invaluable discussions with Professors J. H. Fremlin, Geoffrey Eglinton, Christopher Gregory, Yuval Ne'eman, Roger Penrose, Geoffrey Burbidge, Derek Lawden, Carl Sagan and Freeman J. Dyson. I am especially grateful to the last two for kindly showing me certain papers before their publication.

Grateful thanks are due also to John Delin, Clare Dover, Arthur C. Clarke, Ewan MacNaughton, Kenneth Gatland, Philip Bono, George Hay, Alex Faulkner, Ian Ball, Henry Miller, Enda Jackson, John Anstey, Pat Brangwyn, Harriet Berry, Eleanor Berry, Robin Furneaux, George Herrick, Adrian Fortescue, Yann Weymouth, Michael Pakenham, David Renton, Bob Karbowski, John Glendevon, Leon Jaroff, Virginia Macauley and David Gore-Booth.

I had much useful assistance also from Jeremy Weston and his colleagues at the library at the Royal Institution, London; the staff at the Science Museum Library, Kensington, the London Library, and the National Reference Library for Science and Invention. I am indebted to the public affairs departments of the National Aeronautics and Space Administration at Washington, D.C., and at Houston.

Introduction

This is a remarkable book. When I first read it, in manuscript form, I found some of its concepts staggering; I still do. I have no doubt that some critics will regard the more futuristic ideas as far-fetched. Yet before we voice any scepticism, it is surely wise to look back in time and try to work out what our ancestors would have thought about our own world of A.D. 1973.

Remember, first, that Adrian Berry has called his book *The Next Ten Thousand Years*. It is important to bear the time-scale in mind, because, for one thing, the history of well-documented civilization does not extend back for an equal period. So let us go back a mere 950 years, approximately one-tenth as far. We find that the throne of England is occupied by King Canute, who was a highly enlightened man judged by the standards of his age. Now assume that in the Court there is some Danish or Saxon Adrian Berry, who writes a book entitled *The Next Thousand Years*, and submits it for criticism. How will it be received?

Concept will follow concept. Huge machines, travelling along iron tracks at what would seem fantastic speeds. Pictures sent through the air, so that it is possible to sit in Winchester, turn a knob, look at a screen, and see what is happening thousands of miles away. Flying devices, far exceeding the speed of sound, and capable of travelling from England to the continents over the Atlantic in a matter of hours. And, perhaps more than anything else in the realm of fantasy, spaceships able to send men to the strange, inhospitable world of the Moon. Can we really expect King Canute and his courtiers to accept anything of this? Our hypothetical author will be regarded not merely as an eccentric, but as a lunatic—and probably a blasphemer as well.

Yet in less than a thousand years after Canute, all this has happened; and we must also remember that progress today, from a purely technological point of view, is more rapid than it has ever been. The world of A.D. 1020 was different from the world of A.D. 1820; but the difference between (say) 1820 and 1920 was greater still, and between our present year of 1973 and the end of the twentieth century there will be developments which will cause radical alterations not only in our thought, but also in our way of existence. Of course, we can indulge in speculation, and very often we will be correct, just as a few visionaries (Arthur Clarke in particular) were correct in the 1940s when they claimed that the Moon would be reached before 1970. But as we look further and further ahead, our speculations must inevitably become less reliable, and only the bold spirits among us will dare to make them in serious, scientific vein. Adrian Berry is one of these bold spirits, though he takes care to keep his feet on the ground as much as is reasonable and possible.

Consider, for instance, the section in this book dealing with the practical use of bases set up on the Moon. During the Apollo programme much was heard of the 'Space Is Useless!' brigade, whose members were distinguished as much by their fanaticism and their wish to make themselves heard as for their scientific ignorance. I rather doubt whether anyone who takes the trouble to investigate the problem will maintain that Lunar stations will be anything but beneficial to humanity as a whole. But this is looking ahead only a few decades, not ten thousand years or anything like it; in some degree Lunar bases already exist, even though they are not manned—and we still receive regular reports from the ALSEP stations left on the Moon by the Apollo astronauts, to say nothing of the remarkable Russian automatic devices. There is nothing fantastic about this, and the launching of a new Lunar probe does not nowadays merit even a newspaper headline; yet I can well remember the time when the idea of reaching the Moon was still officially ridiculed.

What next? Look at the remarkable idea of making Venus habitable. At the moment Venus is what Adrian Berry so aptly calls it: a hell-planet, with a fiercely-hot surface and a dense, unbreathable atmosphere. To adapt it to our needs is probably well beyond our present technology; but sending an aircraft direct from London to New York was equally beyond the technology of the Edwardians. The 'Sagan plan' described here may arouse instinctive scepticism because it is unfamiliar—but unfamiliarity should not breed scepticism, just as familiarity should not breed contempt. There may, of course, be some major flaw in the argument given by Sagan and outlined here, but we have no right to assume anything of the kind.

The same is true, to an even greater extent, of the concepts which lie much further from us. Some of them do admittedly sound wild—can we, for instance, make Mars habitable by utilizing the ice locked up in Enceladus, the dim satellite of remote Saturn? And what about the 'Dyson sphere', made up by dismantling some large planetary body and providing the Sun with a sort of artificial blanket? It sounds like science fiction, and at the moment it is something which we can discuss only in vague terms. Yet the concept originates not with a fiction-writer, but with a scientist who has won exceptionally high honours in the scientific communtiy.

The picture of the Earth as an insignificant body in the universe was slow to take root, and it is only in fairly recent times that we have come to accept it; one of Giordano Bruno's heinous crimes, for which he was burned alive in 1600, was that he believed the Earth to go round the Sun instead of vice versa. Whether there are other intelligent races in our own part of the Galaxy must remain a matter for debate—but is it inconceivable that some of them, if they exist, have progressed far beyond our own state, and have even constructed Dyson spheres? This is a question which we cannot yet answer, but it does seem likely that every developing race goes through a period when its technology outstrips its culture. Terrestrial man is in such a state now. As Adrian Berry points out, the savagery of the Neanderthalers and the Cro-Magnons greatly exceeded their nobility, and this is probably the case with ourselves. Can we guarantee that demented social systems will not recur, and engulf not only a few nations but perhaps the entire world? We cannot discount it; but to wipe out all life on Earth permanently would be fortunately rather difficult if we accept the conditions which Adrian Berry lays down. The problem may be with us for a few decades, a few centuries, and perhaps even for the next ten thousand years; but I rather doubt it. By, say, A.D. 2050 we ought to see which turn *homo sapiens* will take.

Here, then, we have a book which surveys many problems and

makes many speculations. You may not credit it all; but read it—
and think hard. I very much hope that some enterprising person
will take a copy, bury it deep in a safe place, and try to ensure that
it has at least a reasonable chance of being dug up and re-read in
the distant future. I wonder what a librarian of the year A.D.
11,973 would make of it? By then, Superspace and Dyson spheres
may be as familiar as trains and aircraft are to us; at least we may
be sure that the Earth will have changed, and Man with it.

And with this, I invite you to join with Adrian Berry as he looks
into the future. I repeat—unfamiliarity should not breed scepticism; and remember that King Canute, a mere 950 years ago,
would never have brought himself to believe in radio, television,
or Alan Shepard driving a golf-ball across the bleak surface of the
Moon.

Selsey   PATRICK MOORE, F.R.A.S.
April 1973

The human mind is not capable of grasping the Universe. We are like a little child entering a huge library. The walls are covered to the ceilings with books in many different tongues. The child knows that someone must have written these books. It does not know who or how. It does not understand the languages in which they are written. But the child notes a definite plan in the arrangement of the books—a mysterious order which it does not comprehend, but only dimly suspects.

ALBERT EINSTEIN

1

The New Atlantis

A study of the future demands some appreciation of the novel philosophy worked out in the seventeenth century by Sir Francis Bacon. Bacon's ideas were quite startling then; yet they lie at the root of modern and future civilization. They effected a scientific revolution as significant as the Industrial Revolution, which followed it a century and a half later and which was its direct consequence. Bacon preached simply that the proper function of science is to search for truth and to benefit people. This idea, which is commonplace to us, was almost unheard of in his time. Before the age of Bacon, science, or natural philosophy as it was then called, was considered to have a quite different purpose. Any belief that science and technology were connected, or ought to be connected, was accounted both vulgar and absurd. All natural philosophy was undertaken by the 'schoolmen', those hair-splitting thinkers who reasoned in the classical style of Seneca, the most famous of the Stoics, and above all of Plato.

Before the time of Bacon, there had indeed been some scientific progress. Galileo had observed through his primitive telescope that the planet Jupiter had its own moons which revolved around itself and not around the Earth; Copernicus had shown that the Sun was the true centre of the Solar System. But these curious items of information were not considered by the educated public to be true 'science'. Far more fundamental questions occupied the most brilliant and disputatious minds of previous centuries. What is the greatest good? Is pain good or evil? Are all events predestined? Can we be certain that we are certain of nothing? Can a wise man be unhappy? Are all departures from right equally reprehensible? How many angels can stand on the point of a needle? Generations of learned men argued these questions, and, reaching no verifiable conclusions, added little to the stock of human knowledge. Abelard, the most respected philosopher of the twelfth century, was typical of his time when he announced that dialectic is the sole road to truth.[1] John of Salisbury, secretary to Archbishop Thomas Becket, revisited one school of philosophy, and was astonished to find them still debating the same problems as they had been thirty years before.[2] Round and round the schoolmen argued. In the words of one historian, 'they filled the world with long beards and long words, and they left it as ignorant as they found it.'[3]

All scientific speculation before Bacon was restricted by the classical principles. Far from enlightening mankind or being of any service to him, these concepts obstructed the progress of knowledge for nearly two millennia. Posidonius, a distinguished writer in Julius Caesar's time, ventured in one essay to cite the principle of the arch and the use of metals as some of the humbler blessings which man owed to natural philosophy. Seneca repudiated these compliments, which he regarded as an insult. Natural philosophy, he declared, had nothing to do with teaching men to erect arches over their heads, nor with the use of metals.[4] The true philosopher does not care whether he has an arch over his head, or whether he is exposed to the wind and rain. Far from exploring the uses of metals, thundered Seneca, philosophy teaches us to live without materials and mechanics. One passage from Seneca expresses precisely the pre-Bacon attitude towards science:

> In my own time there have been inventions of this sort, transparent windows, tubes for diffusing warmth equally through all parts of a building, short-hand, which has been carried to such a perfection that a writer can keep pace with the most rapid speaker. But the inventing of such things is drudgery for the lowest slaves; philisophy lies deeper. It is not her office to teach men how to use their hands. The object of her lessons is to form the soul. Next we shall be told that the first shoemaker was a philosopher![5]

These sentiments, which prevailed among nearly all intellectual and religious authorities until Bacon, were strongly influenced by Aristotle* and were supported by Plato and Socrates. As Bertrand

* Aristotle's metaphysical and mystical writings have tended to obscure his genuinely scientific books, such as *The Politics* and *The History of Animals*. The former was an accurate catalogue of all possible political

Russell remarks, 'the ethical and aesthetic bias of Aristotle, and still more of Plato, did much to kill Greek science.'[6] Even Archimedes, the greatest engineer of classical times, who held a Roman fleet at bay for three years with his lethal mechanical inventions, felt somewhat guilty about these works, regarding them as little more than relaxation from the rigours of high-minded philosophy.[7]

Francis Bacon left Trinity College, Cambridge, at the age of fifteen, with a deep contempt for what he considered to be its narrow-minded dialectic teachings under the direction of its master, that bigoted persecutor John Whitgift. Even at that age he is said to have been meditating the great scheme of philosophical reform which has carried his fame through the centuries. Whitgift's teachings had on Bacon the opposite effect to that intended. The boy conceived a hatred for the Platonist and Aristotelian schools of philosophy which grew with the maturing of his mind. He decided to imitate Plato in setting down his vision of a Utopia. Both Utopias, Plato's Republic and Bacon's New Atlantis, resembled each other in that they represented the ideal worlds of these two great men. Yet how they differed! Plato desired a rigid, authoritarian world, somewhat modelled on Sparta, in which each citizen would be kept in his allotted place. Bacon, inspired by the aggressive capitalism of Elizabethan England, imagined a commercially and scientifically oriented community where everyone would benefit from the successes of entrepreneurs.

It is interesting to compare in detail some of the differences between these two philosophies. To Plato, the modern approach to astronomy would have been odious. He declared that to use astronomy for determining the seasons, for agriculture or navigation, or for any other practical purpose, or even for seeking to understand the Universe, was a vulgar activity, unworthy of philosophy. Knowledge of the motion of heavenly bodies was to him of no value. The shapes of the constellations were mere examples, mere aids to feeble minds: 'We must get beyond them: we must neglect them. We must attain to an astronomy which is as independent of the actual stars as geometrical truth is indepen-

---

systems. His observations of marine life in the latter have been confirmed by modern biologists.

dent of an ill-drawn diagram . . . The true purpose of astronomy is not to add to the vulgar comforts of life, but to raise the mind to the contemplation of things which can be perceived by pure intellect alone.'[8] This kind of language was incomprehensible to Bacon, to whom astronomy was only useful when it concerned itself with truth or profit. He compared the astronomy of Plato with the ox of Prometheus, 'a sleek, well-shaped hide, goodly to look at, but stuffed with rubbish and containing nothing to eat.'[9]

Plato finds no great merit in the invention of handwriting, because it teaches men to be idle. By keeping books, without having to memorize their contents, a man can have information whenever he wants it. Such a man cannot strictly be said to *know* anything. All books, Plato declares, should be discarded as soon as they have been read. Their owner would then be compelled to develop his memory, and by deep meditation heighten the powers of his intellect.[10] But Bacon, as may be imagined, cannot see in the least why unnecessary feats of memory should be any use to a man. He compares them with exhibitions of rope-dancing and juggling. 'The two performances', he says, 'are much of the same sort. One is an abuse of the mind; the other is an abuse of the body. Both may excite our wonder, but neither is entitled to our respect.'[11]

Plato is equally contemptuous about medicine. He has no patience with invalids. A life protracted by medical skill is to him a long death. Those of frail health should be allowed to die. Such people are unfit for war, statesmanship or severe study. If they engage in mental exercise they become giddy in the head and interrupt serious discussion. Plato does not object to surgery for a healthy person wounded in battle or accident, but for sufferers from degenerative diseases he has no sympathy. Are not such illnesses the natural and justified punishment for gluttony and lust?[12]

Bacon has no time for such harshness. He holds that sick people should be enabled to end their lives in comfort rather than in pain. While being uninterested himself in studying medicine, he urges that research should be bent to find means of mitigating the symptoms of degenerative diseases. Since Plato had cited characters from Homer to bolster his statements, Bacon retaliates by enumerating the healing miracles of Christ.[13]

The severe philosophies of Plato and Seneca held sway for long ages, during which the genius of man was sidetracked into a false direction. Poverty was noble; luxury was evil. Scientific speculation was only to be tolerated so long as its function was to sharpen the wits and not to improve material comforts, discover truth or make money. Brilliant disputation was the most excellent of activities. Above all, it must not be sullied by materialism. Even these ideals became perverted in the Middle Ages as the remembered teachings of the ancients became blurred. Luxury and hypocrisy flourished. These, except in the minority view of Dante,[14] were not considered to be great sins. What a man did was less important than what he said. There was a curious marriage between classical philosophy and the most intolerant dicta from the Old Testament. Leaders of the Church might array themselves with every sort of luxury and wealth, but so long as they preached theological purity, they could maintain para-military Inquisitions in the lands of devout kings whose armies could have crushed them within a day. George Orwell's discovery of double-think seems as applicable to the Middle Ages as to our own times, whose vices he mocked.

This general state of mind explains many things; why the kidnapped Inca of Peru was lectured on the blessings of poverty by men determined to steal his wealth in order to acquire it for themselves.[15] It explains why Galileo was condemned by a caucus of churchmen who thought him a vulgar blasphemer. And it explains why Martin Luther, who knew nothing of mathematics or astronomy, thought fit to declare that Copernicus was a fool, and that his theories about the Solar System were anti-Biblical and intolerable.[16] It was seldom reflected that Seneca, one of the most eloquent promoters of this repressive creed, was a man of such grave faults of character that his teachings might be of questionable value. He wrote a pamphlet on virtue, within days of writing another to justify Nero's murder of his mother. He praised the virtues of poverty, with millions out at usury. He denounced the evils of luxury, while taking his ease in the gardens of Lucullus.[17]

If the personal vices of a philosopher are to be taken into account when assessing the value of his philosophy, it must be admitted in fairness that Bacon also had a sordid political career.[18]

To please Queen Elizabeth I, he laboured to send to the scaffold the Earl of Essex, the man who had spent much of his life seeking lucrative posts for his friend Bacon. Bacon took a leading part in the judicial murder of that bold explorer Sir Walter Raleigh. Acquiring high office under James I, he supervised the torture of a nonconformist clergyman on account of some unpreached sermons found in the wretched man's drawer. In 1621, he was dismissed from the post of Lord Chancellor after admitting to both Houses of Parliament that he had accepted bribes. Yet unlike Seneca his contributions to civilization infinitely outweigh his deficiencies as a man. After 1,500 years the teachings of Seneca and the ancients have left us nothing but theological disputes with no solutions and no meanings. The scientific achievements before Bacon's time were made despite classical philosophy, and, as we have seen, were much restricted by it. But 300 years after Bacon published his writings, and as a direct result of them, we have the marvellous conveniences of modern life; electronics, nuclear energy, computers, super highways, jet aircraft, space travel, and the richness and splendour of modern astronomy and sub-atomic physics. We can measure the smallest particle, cure some of the most serious diseases, cross vast oceans within hours, and peer at the edge of the Universe—all because a corrupt politician declared three centuries ago that the search for truth and usefulness was more profitable than disputations about the meanings of words.

Bacon himself knew nothing of any of these things. If he were miraculously deposited in the 1970s, he would be as astonished as any of his contemporaries. But he is regarded as the originator of the maxim 'knowledge is power'.[19] He saw clearly *how* technology could be achieved, and the benefits which it must surely bring to man. It was his Utopia, *The New Atlantis*, published in 1627, which created the most profound impression. This was a novel, somewhat in the style which Swift later used for the adventures of Gulliver. The work contained a complete description of how science and technology should be organized for the good of the community and the enrichment of its members.[20] Bacon's hero sails across an ocean and discovers the fabulous kingdom of Bensalam. This is no comic satire on English politics, like Swift's Lilliput and Brobdingnag. Bensalem bears not the faintest resemblance to Bacon's England; it is more comparable to modern Japan. The citizens of Bensalem are enormously advanced. They have aircraft and submarines, refrigerators and hearing aids. These are some of the fruits of a brilliant social and economic organization. The most important place in Bensalem is the House of Solomon, which acts as a clearing house for all industrial and scientific information. The House of Solomon employs a body of men called the Merchants of Light, who play the role of industrial spies. They travel secretly to foreign countries, collecting all information which might be useful to Bensalem. The words 'profitable inventions' recur frequently in the novel. The entire political structure of Bensalem is geared to the premise that the noblest thing a man can do is to invent something that will enrich his shareholders and raise the living standards of his customers. It is a sort of nationalistic capitalism, aimed at ever increasing the happiness and prosperity of the individual through the exploitation of science. The information acquired by the Merchants of Light is tested and analysed by the Depredators. These, in turn, suggest practical applications to the Benefactors, who report on them to the commercial companies, after 'casting about how to draw out of them things of use and practice for man's life and knowledge'. But it is not the companies who govern the country. The real rulers are the pure scientists, the Interpreters of Nature, who decide upon new fields of research, who direct the Merchants of Light, and who order former discoveries 'to be raised by experiments into greater observations, axioms and aphorisms'.

It is hard for us to imagine the terrific impression which Bacon's ideas made upon seventeenth-century minds. It was perceived for the first time that humanity might have a hidden purpose, and might be able to execute a long-term plan whose nature had been hitherto concealed. The notion that science could be made to work more profitably if research was systematically organized and freed from religious interference was completely new. These ideas were soon put into execution. In 1660, scarcely more than a generation after the publication of *The New Atlantis*, The Royal Society was formed, to put into practice the scientific methods of Bensalem. The Society states in its *Record* that its foundation was 'one of the earliest practical fruits of the philosophical labours of Francis

Bacon'. The poet Abraham Cowley wrote an ode to the Royal Society, declaring:

> Bacon, like Moses, led us forth at last,
> The barren wilderness he passed,
> Did on the very border stand,
> Of the blessed promised land,
> And from the mountain top of his exalted wit
> Saw it himself, and showed us it.[21]

King Charles II, who was as enthusiastic about science as his grandfather James I had been dogmatic and superstitious, gave the Society its charter and attended some of its meetings. Sir Isaac Newton became its second president and its most famous member as, with his theory of universal gravitation, he laid the groundwork for modern physics. The prospects for mankind seemed immeasurably improved. The educated men of 1700 were a different breed from those of 1600. In 1600, for example, comets were regarded as portents, as in Shakespeare's *Julius Caesar*:

> When beggars die there are no comets seen:
> The heavens themselves blaze forth the death of princes.

Few dramatists would have dreamed of penning such lines in 1700. By then, comets were no longer thought of as omens, but as tiny planets obeying the universal laws of gravitation. In 1700 the outlook of educated people was largely modern. A century earlier, even allowing for such adventurers as Francis Drake and John Dee, it was in general medieval. This great change appears to have been principally due to Bacon. All within a century, there came a series of advances which might never have been possible, or at least would have been much delayed, without his system of scientific organization and the communication through journals of theories and the results of experiments. From Thomas Sydenham in 1666 came the introduction of clinical medicine. Christian Huygens in 1678 developed the wave theory of light. Newton, in 1693, followed up his gravitational theory with the invention of infinitesimal calculus. The eighteenth century brought the modern theory of combustion and the law of conservation of matter from Lavoisier, and the technique of smallpox vaccination from Edward Jenner. *The New Atlantis* reappeared in four successive editions. And the most important new development of Bacon's philosophy came in 1751 when Diderot and d'Alembert published the first volume of the immensely influential *Encyclopedia*, that vast collection of all known scientific facts and theories which directly preceded the Industrial Revolution. The two editors announced their desire to further Bacon's work. 'We are tempted to esteem him the greatest, the most universal and most eloquent of all philosophers,' they said. 'It is to this great author that we are chiefly indebted for our Encyclopedic Plan.'[22]

The revolution wrought by Bacon must be seen as the single most important 'revolution' that has ever occurred in human society. With several volumes of essays and a novel which I have very imperfectly summarized, he set in motion a chain of events, which today has gathered an inexorable momentum. Wars and economic recessions cannot for long turn it aside. He has set us on a course from which there is no retreat; even if we wished to stop all economic growth, as some of the environmentalists urge us to do, there would be nothing to return to but poverty, disease and urban squalor.

Yet if Bacon, or someone like him, had never existed, the world might today be scarcely more advanced than when he was born. We might still be ensnared in the unpractical philosophies of Seneca and Plato. The roots of these two opposing philosophies are easily discerned. The historian Macaulay in 1837 composed an amusing fantasy in which Seneca and Bacon meet as travellers.[23] They reach a town where smallpox is raging. All houses are shut, trade is suspended, and mothers weep in terror over their children. Seneca delivers a lecture on the nobility of suffering. Bacon takes out a lancet and begins to vaccinate. They then meet a group of miners who are in great dismay because an underground explosion has trapped many of their comrades. Seneca urges them to turn this tragedy to their intellectual advantage by using it as an example to show that all events, good and bad, are equidistant from eternity. Bacon, who has no such fine phrases at his command, designs a safety lamp and rescues some of the trapped men. On a seashore they meet a shipwrecked merchant wringing his hands. His valuable cargo has gone down, and he is reduced in a moment from opulence to beggary. Seneca exhorts him to seek happiness in things that lie outside himself. Bacon constructs a diving bell and returns with the most precious objects from the wreck. Fables like these could be invented endlessly to show the essential differences between the ancient philosophy of words and the modern philosophy of works.*

Bacon's revolution is today only in its infancy. Modern technology appears marvellously advanced—but only when it is compared with the technology of a hundred years ago. Some centuries from now, our most sophisticated computers will be but scrap metal and our fastest aircraft will attract the same amused interest as does a dugout canoe in a museum of anthropology. The thesis of this book is that economic progress and technology are going to continue, not merely for decades, not even for centuries, but for millennia. Earth cannot provide the living space and the raw materials for such colossal geometric progression; space itself will be exploited. The planets around the Sun will be inhabited and industrialized. Jupiter, the largest of these planets, will be dismantled and its fragments displaced to capture the Sun's radiation more efficiently.

But even these great projects will be a mere beginning. Beyond the farthest planet of this local Solar System lies our great Galaxy of stars. Among them is man's true destiny during the next 10,000 years. No earthly array of jewels can compare in splendour with colour photographs of the galactic star fields taken through a big telescope. The brilliant blue of the giant stars, the white and yellow stars of similar size to our own Sun, the blood-red dying suns, all dazzle the viewer from behind the veil-like orange nebulae of clouds of dust and hydrogen from which new stars will be born. Here is the hidden reality of Bacon's *New Atlantis*! The human exploration and colonization of this almost unbounded expanse of suns with their accompanying planets represents the true future activity of man. It will be shown in this book how it may prove possible to traverse interstellar distances at much greater speeds than present interpretations of physical laws suggest. What therefore must be imagined? Not a planetary civilization restricted to a single world, not even a community of Solar worlds of static progress, cringing from the immensities of interstellar space, in the way that the Romans feared the terrifying vastness of the Atlantic

---

* One is reminded of the Royal Society's motto, *nullius in verba* (words alone are nothing).

Ocean. Rather we should foresee a series of human empires in the Galaxy, dominions perhaps of one planetary system over millions of others. Economic, social and psychological necessity will drive us to this course.

The Russian astronomer N. S. Kardashev, when speculating about intelligent civilizations in the Universe, predicted that a planetary civilization will go through three general phases.[24] The modern world, according to Kardashev, is a Phase 1 civilization, which draws upon the energy resources of a single planet. Phase 2 occurs as the entire parent Solar System is utilized, with the giant planets being dismantled and displaced to use their raw materials. But even this is modest compared with Phase 3, in which whole sections *of a galaxy* are exploited. Kardashev's prophecy seems applicable to ourselves. The Sun has scarcely reached the half-way point of its stable life, and the Galaxy is expected to endure far more than ten times its present age. As the great astronomer Sir James Jeans wrote in 1930:

> We are living at the very beginning of time. We have come into being in the fresh glory of the dawn, and a day of almost unthinkable length stretches before us with unimaginable opportunities for accomplishment. Our descendants of far-off ages, looking down this long vista of time from the other end, will see our present age as the misty morning of human history. Our contemporaries of today will appear as dim, heroic figures who fought their way through jungles of ignorance, error and superstition to discover truth.[25]

2

No Limits to Growth

Far from taking us to the stars, might not Bacon's policy be leading us to disaster? A literary journal says that Bacon has bequeathed to us every sort of 'technological horror'.[1] Fears have been widely expressed since 1968 that ever-increasing economic and population growth, with consequent increase in pollution, will, before another century has elapsed, bring about the collapse of civilization. Why has this tendency in human thought become especially strong since 1968? That was the year of Apollo 8, the first manned space-flight round the Moon, when for the first time it was possible for man to look through a porthole and see the entire Earth as a sphere. Sir Fred Hoyle and others have suggested that concern about the Earth's environment was enormously heightened by the colour photographs from this mission.[2] The bright blue Earth seemed to hang in space like an orbiting ship. It appeared lonely and defenceless in the blackness. As the poet Archibald Macleish wrote, describing the view from Apollo 8:

> To see the Earth as it truly is
> Is to see ourselves as riders on the Earth together,
> Brothers on that bright loveliness in the eternal cold.

Whatever the psychological effects of these pictures may have been, it was about this time that environmentalists began talking with great stridency about 'Spaceship Earth'. Books began to appear with alarming titles, like *The Doomsday Book*, *The Population Bomb*, *Only One Earth*, *Eco-Doom*, and *Famine—1975!* Journals were launched with names like the *Ecologist*. A group was formed which called itself Friends of the Earth.*

The climax of these efforts to persuade the public that man had

---

* Dr Edward Teller, on hearing of the existence of an organization called Friends of the Earth, remarked, 'Perhaps the Earth has too many friends and the energy-user too few.'

a doubtful future came in 1972 with *The Limits to Growth*, published by a body of scientists and industrialists known as the Club of Rome. This book, which has sold several hundred thousand copies to date, has gained a dual reputation for the promise of its methodology and the falsity of its data and conclusions. The authors claimed to have assembled, in mathematical form, all known data about world population, food supplies, pollution, and industrial raw materials. They produced a series of computer 'models' purporting to show how each of these variables was going to rise and fall during the next 100 years. They concluded that present trends would assuredly bring about the collapse of civilization before the year 2100, and that the only way to avoid this disaster was to make drastic restrictions to economic and technological growth.

The most admirable part of this book was its mathematical approach. The authors used a method called 'systems dynamics' in which the complicated interaction of variables can be predicted by a computer. This method had been proposed in 1953 by the computer pioneer John von Neumann (see Chapter 11) and apparently used for the first time in 1961 by Jay W. Forrester at the Massachusetts Institute of Technology. But the Club of Rome's book (or 'report' as they preferred to call it) was not a scientific success. Indeed, it had the effect of reminding us that the technique of von Neumann and Forrester is still very much in its infancy, and infants, as a rule, cannot tell us very much that is of value. John Maddox, then editor of the prestigious journal *Nature*, was perhaps somewhat harsh in calling the book 'sinister', and Gunnar Myrdal of Sweden might have been a bit severe in dismissing it as 'pretentious nonsense', but it would be wrong to say that its conclusions gave us anything other than falsehoods.

The data which the Club fed into its computers might have been calculated to bring disgrace upon the whole concept of computer models.[3] 'Pollution' was represented in each model as a single variable. How was this pollution measured? By the density of smog over Los Angeles? By the number of discarded beer-cans found each year in a given district? By the surface area of marine oil-slicks in latitude and longitude such-and-such? The Club gave no definition. Critics asked how 'pollution' could be given a numerical value in the future if no such generally recognized number existed in the present. 'Who can say', wrote three Dutch scientists in an article in *Nature*, 'what the height of the 1970 global pollution level might be in relation to the 1900 level?'[4] Far from recognizing the measures which governments have successfully been taking for years to clean up cities, air and rivers, the Club's models appeared to deny that any such measures had been taken at all. Instead, they predicted that pollution controls would only be made possible 'by high levels of capital investment'. This is the opposite of what has in fact happened in the post-war world. Controls have been imposed, not at a high level of capital investment, but when the level of pollution itself has become too high. An enormous amount has already been achieved. In many places clean-air laws have removed much of the poisonous city smogs that were so notorious earlier this century. Restrictions on sewage have cleaned many rivers to the extent that fish can be seen in them for the first time in fifty years, even in those stretches which run through cities. Even if the carbon monoxide-producing petrol engine is not replaced altogether within the next two decades, as seems very likely, it appears certain that all road vehicle exhausts will have emission controls.

'Resources', like pollution, are lumped together by the Club in a single undefined variable. Their prediction of a 'decline' in resources will certainly be turned into nonsense by future developments. It is no more valid today to glance at the world's currently usable resources and to say, 'They will all be gone by the year such-and-such; so we must hoard them or we are doomed,' than it would have been in 1900. A Club of Rome model written in 1900 would surely have predicted that civilization would collapse long before 1976.* The use of uranium was then unknown, as was nuclear energy. So also were plastics, stainless steel and the use of titanium, and such techniques as welding and superconductivity. The amount of resources available today cannot by themselves give us very much clue about the amount and nature of resources that will be available and useful in the future. Judging by the past, they will comprise a host of materials that are today unknown, or

* An amusing article in the *Economist* (March 11th, 1972) pointed out that any study of London's transport made in 1872 would have predicted that by 1972 the city would surely be buried under a mountain of horse manure.

are still useless laboratory curiosities. The Club of Rome, by contrast, does a disservice to science by asserting that they will be exactly the same as today. However tolerantly we examine the models, we find flaws of logic that would be obvious to every reader if the book's mathematical jargon and technical format did not give it such an air of authority. One model, for instance, shows an unexplained and incredible rise in food consumption per capita, as if the whole world had decided simultaneously to commit suicide by overeating.[5] As Sir Eric Ashby, chairman of Britain's Royal Commission on Environmental Pollution, remarked, 'if you feed doom-laden assumptions into computers, it is not surprising that they predict doom.'[6]

If any suspicion remains that the Club's models might have some sort of validity, we may turn to a paper on the Club of Rome published later in 1972 by the World Bank. Economists often see more deeply into human affairs than ecologists, and this instance was no exception. The Bank's paper reads like an unfavourable school report.[7] Many of the assumptions fed into the models were 'extremely pessimistic' and 'not scientifically established', and the use of data was often 'careless and casual'. The Club had taken statistics about mineral resources from documents published by the U.S. Bureau of Mines, but these figures had been used 'indiscriminately', and 'the small print has been ignored'. Extrapolations into the future had been made 'heavily and dogmatically'. The Club had 'allowed for' current uncertainty about the limits of resources simply by multiplying all proved reserves by five, a procedure which was 'patently absurd'. The Club's claim that the mathematical elegance of its models was a better guide to the future than was past history was 'grossly erroneous'. There was 'absolutely no scientific evidence' for the Club's somewhat mystical announcement that the world's capacity to absorb pollution was exactly four times the amount of pollution now produced. And perhaps worst of all, the Club had sought direct relationships between pairs of its dubious variables, and had then used these alleged relationships 'as if they were natural laws'.

The art of invective is evidently flourishing, but the science of futurology might appear from all this to be in a sick condition. Is there not a more reliable chart or equation, based on data which are not so easily disputed, which can point out to us more accurately the direction of human destiny? Happily there is. We have had since 1913 a remarkable astronomical document known as the Hertzsprung-Russell Diagram. First drawn in that year by Einar Hertzsprung of Denmark and Henry Norris Russell of the United States, it has since been revised and redrawn many times. It is a statistical chart of several million comparatively close stars, showing the relationship between their colours, their absolute magnitude of brightness* and their spectral types. From this information can be calculated their mass and the amount of time for which they are going to remain stable. What has all this to do with the destiny of man? The Diagram tells us a great deal about one star which is at present the most important to us—the Sun. Our Sun is situated almost at the centre of the curving diagonal strip which goes from the Diagram's top left corner to its bottom right. This strip represents the 'main sequence', or those stars which are presumed to have been stable, with no noticeable change in brightness or volume, since their creation.

Mass (Sun = 1)	Time on the main sequence (millions of years)
30	0·05
15	150
1·5	2,100
1·4	3,500
1·3	4,900
1·2	6,500
1·1	8,300
1·0	11,000
0·5	100,000

Now, the Sun's mass is well known, and we can compare it with the masses of many other stars. A pattern of probable behaviour soon appears. In general, the rule is this; the more massive the star the shorter its life on the main sequence. A star leaves the main sequence when its hydrogen fuel is exhausted. It then either explodes, if it is very massive; but more usually it swells into a red

---

* I.e. their real brightness, as opposed to their 'apparent' brightness when seen from Earth. The absolute brightness of stars is measured as if they were all at a theoretical distance of 32·6 light-years from Earth.

giant, increasing its diameter about thirty times. Either of these forms of stellar death will obviously obliterate most of the planets in orbit around the stars. Even if its planetary companions are not reduced to gas, any life on them will at once come to an end. When will this happen to the Sun and its planets? We consult the Diagram, and the answer makes us thankful that the Sun is no more massive than it is. The table opposite shows the approximate lifetime on the main sequence of stars of varying masses.[8] The average margin of error of time-scales is about 10 per cent. In short, the lifetime of the Sun, from birth to red-giant stage, will total about 11 billion years.* It is generally accepted that some 5 billion of these years have already elapsed. We can say therefore that the Sun will continue to shine at roughly its present strength for another 6 billion years. The level of Solar radiation may vary on occasions, but judging by the Earth's known history and the examples of other main sequence stars in the Diagram, these variations will never be radical enough to endanger the planet. Ice ages, whether or not they prove to be induced by Solar radiation changes, may come and go; but they have not so far brought about a total cessation of life, and there is no reason to fear that they will do so in the future. Some malevolent person may succeed in blowing up the Sun, but barring this the Earth, and all the life it carries with it, is here for the duration.

Six billion years may seem a truly fantastic period of time, almost beyond the imagination of many people. But the Diagram is really telling us that *six thousand thousand thousand* years of history are going to pass on Earth before nature rings down the curtain.[9] This sort of time-scale makes fears about the long-term effects of a nuclear war or of other disasters seem rather ridiculous. The 2 million-year time-scale of human evolution is a mere fraction of a moment when measured against 6 billion years. (It is one three-thousandth, to be exact.) Too many books about space travel and other futuristic projects, when promising great achievements, have added the caution, 'provided we don't blow ourselves up'. It never seems to occur to these writers that a large number of people are certain to survive the 'blowing up'. It is true that nuclear weapons have been stockpiled to such an extent that the

* The 'billions' thoughout this book are American, i.e. thousands of millions.

equivalent in explosive power of 10 tons of T.N.T. exists for every human being in the world. But that is a very different thing from saying that in a nuclear war each human being would be duly killed by his 10 tons of T.N.T. The megatons of explosive power are united in single bombs, and cannot be scattered uniformly over the globe. The worst imaginable nuclear war was postulated by two U.S. Congressional committees, chaired by Congressman Chet Holifeld, during 1958, 1960 and 1961. After hearing innumerable expert witnesses, they announced the casualties of their 'war'.[10]

The exercise told us a great deal about the consequences of a nuclear war. In a long-term sense the results were reassuring; but for those thinking in terms of years rather than centuries they were very horrible. Holifield's team assumed that two 10-megaton bombs were dropped on New York City, where a 40-m.p.h. wind was blowing west-south-west, carrying the fallout towards the most densely populated areas. Bombs totalling 275 megatons exploded in the area between Boston and Washington, and in all some 1,450 megatons fell throughout the United States. Thirty-one per cent of the population (about 56 million people) were killed at once. Another 12 per cent (about 22 million people) suffered injuries. About 21 million of the country's 46 million houses were 'moderately or severely damaged'. A further half-million were so badly affected by fallout that they would have to be evacuated for at least a year. Fuel and transport services were temporarily wrecked.

Inverting these figures, we see that there would be strong grounds for expecting a complete recovery within decades. Fifty-seven per cent of the population, or 103 million people, were unhurt in Holifield's war. Twenty-four and a half million houses were free both from fall-out and from blast damage. All the belligerent countries would suffer similar losses—except for very densely populated industrial countries like Britain and Holland, where the proportion of casualties would be much higher.

Holifield's study was followed a year later by a more optimistic book, Herman Kahn's famous *On Thermonuclear War*. Kahn drew scenarios of several nuclear wars of varying intensity.* He

* Kahn was bitterly criticized for writing this book, not because his figures were disputable, but because some people thought it repulsive that

concluded that casualties would be as bad as, if not worse than, Holifield's committees had feared. But I call his book 'optimistic' because, instead of lamenting the dead and dying, he concentrated on the prospects for the healthy survivors. After describing each imaginary nuclear war, he asked, 'Will the survivors envy the dead?' In each case, his answer was 'No'. After innumerable interviews with economists, agricultural geneticists and other specialists, he estimated that it would take no more than 40 years for the United States to recover to the condition in which it had been before the attack. Only 40 years? Suppose that Kahn's estimation was far too optimistic, and that he erred by a factor of 10 or even 100. It would then take the United States and other attacked countries 400 or 4,000 years to recover. These would indeed be long dark ages to the generations that lived through them, but when measured against the time-span predicted for us by the Hertzsprung-Russell Diagram, they are nothing. A nuclear war would be a horrible disaster, but when considered on the scale of the Earth's future history it would not matter in the slightest. There is, moreover, no reason to fear that Kahn's estimate of 40 years was wrong. To make a parallel on a much smaller scale, we saw how quickly Coventry, Warsaw and Dresden recovered after conventional saturation bombing, and how completely Nagasaki and Hiroshima appear to have recovered from the atomic attacks which seemed at the time to have utterly destroyed them.

The complete destruction of the human race appears an almost impossible task. To make intelligent life *permanently* extinct on this planet, it would be necessary to:

1. Kill every human being; if a single group of so much as 500 people of both sexes survived in one place, the operation would be futile. Our civilization could then regain its former state in as little as half a million years.

2. Kill all the apes and monkeys throughout the world. Branches from any one of their species could in time (a few million years) evolve into a powerful technological civilization.

3. Kill all squirrels, tree-shrews and all other tree-dwelling

nuclear war should be written about in a cool mathematical tone. A review in *Scientific American* (March, 1961) called the book a 'devilish, blasphemous work'. Kahn wrote indignantly to the editor: 'Your magazine is neither scientific nor American.'

mammals. Our own ancestors are believed to have been animals of this kind some 70 million years ago.

4. Destroy all trees and all plant life and somehow stagnate the oceans to deprive any surviving species of oxygen.

5. Repeat the last operation every million years or so. Once plant life had re-established itself, a life-giving oxygen atmosphere would soon follow.

In the long run, therefore, the world appears to be almost indestructible as a habitat for life for a very long time. Provided he controls his numbers, it seems highly probable that man has an age-long future. The world population is at present growing at about 2 per cent each year. The statistics are well known to any-one who reads environmental literature; if continued, this rate of increase would mean a doubling every 35 years, raising our present population of 3·7 billion in 1972 to 6·5 billion in the year 2000, to 47 billion in 2100, and 1,000 years from now, in the year 3000, we would reach the fantastic total of 2·5 *quadrillion* (i.e. 2·5 million million million). Even if all the oceans were evaporated to provide more living space, this would still mean that thousands of people were occupying each square yard. The proposition is plainly absurd, and the growth rate is bound to slacken and reach some kind of stability. The question is, when? The prospect of this happening too slowly would be terrifying—if it were more convincing. Paul Ehrlich began his campaigning book *The Population Bomb* with the following anecdote:

> I came to understand the population explosion emotionally one stinking hot night in Delhi a few years ago. My wife and daughter and I were returning to our hotel in an ancient taxi. We entered a crowded slum area. The streets seemed alive with people. People eating, people washing, people sleeping, people visiting, arguing and screaming. People thrusting their hands through the taxi window, begging. People defecating and urinating. People clinging to buses. People herding animals. People, people, people, people. Since that night I've known the *feel* of overpopulation.[11]

This is the kind of future for the whole habitable world of which Ehrlich goes on to warn us. Similar predictions of doom come from William and Paul Paddock with their *Famine—1975!*[12] It should be said at once that despite Ehrlich's exciting prose quoted above, these are both very bad books. Their statistics, like those of the Club of Rome, are based on wrong assumptions; facts which contradict their theses have been ignored. The international policies which they urge for reducing the world population growth would, if adopted, assuredly have the effect of accelerating it. Ehrlich and the Paddock brothers have had great influence on the environmental movement, and one has a nagging anxiety that their arguments might, in an unlucky moment, influence political decision-makers. They adopt the same position as did Thomas Malthus in 1798, when he published the first edition of his famous Essay on the Principle of Population, which urged that population would increase until checked by famine, plague and war. It is not often remembered, certainly not by Ehrlich and the Paddocks, that Malthus's second edition of this essay, published in 1803, contained the admission that population could be controlled by what he called 'moral restraint', and what we might call 'prudence'. What did he mean?

Some researchers have found that animals instinctively impose birth control upon themselves. One of these specialists, the American John Calhoun, did an important experiment with wild Norway rats in a large pen which could have accommodated 5,000 healthy rats. He fed them regularly, and saw no reason at first why, within two years, they should not reach their maximum theoretical population of 50,000. Yet having started with five rats, the population reached 200 and then stopped. Every indication suggested that no matter how long the experiment continued this would be their approximate number.[13] Calhoun's experiment has never been quoted by the population alarmists, as far as I know, perhaps because it runs counter to their warnings.

The statisticians of the United Nations also have discovered what Malthus was implying by 'moral restraint', although he himself only defined this concept vaguely. The United Nations is at present the least unreliable source for world population statistics, even if, as they admit, their figures are subject to fair margins of error. A surprising pattern emerges. Contrary to all expectations, it appears that the availability of medicine is the most powerful regulator of population growth. It may be no coincidence that in the 'developed' world, e.g. the United States, Russia,

Europe and Japan, where medical techniques are advanced, and where infant mortality and the proportion of deaths by disease and starvation is low, the population is growing at less than 1 per cent. In the 'developing' world, by contrast, we find growths varying between 2 and 4 per cent. The developed world has virtually reached demographic stability, and some of the undeveloped nations are moving towards it. It appears that high and potentially disastrous growth-rates occur when parents expect that most of their children are going to die. Yet the parents are often too pessimistic, and it chances that all their children survive. But it is too late; they have already allowed for the possibility of the deaths of their children by having several more. An enormously high birth-rate results. Now introduce modern medicine to a developing country. Within a few years there is a change. Parents no longer fear the deaths of all their children, and no longer feel that they must insure against this possibility by having great numbers of them. A decline in the death-rate is soon followed by a decline in the birth-rate. Marriages are delayed, the growth-rate slackens, and population stability is gradually achieved. This is exactly what has happened in the developed countries, and it is beginning to happen in the rest of the world. There is a peak year of very high birth-rate; then medical technology improves, and births decline. Opposite are some U.N. figures of declining births in selected countries which clearly illustrate this trend. [14]

In a large number of developing countries, therefore, the birth-rate is falling substantially and it seems very unlikely that the 2 per cent growth-rate of population will be maintained for the rest of this century, even if it has not been reduced already.* And this slackening, as we have seen, is largely due to improved medicine. But what solution do Ehrlich and the Paddock brothers advocate for our population problem? They suggest, quite seriously, that American foreign aid should be denied to those countries which do not have vigorous birth-control programmes. In other words, starve them of medical aid (Ehrlich calls it 'exported death con-

---

* All population statistics are little more than rough estimates. Previous figures for the British population, for instance, were found by the 1971 census to be at least 1 per cent in error. Only 66 of the 132 countries in the United Nations have returned reasonably accurate data for two recent years (1966 and 1970). It is possible that our present 3·6 billion estimate for the world may be in error by as much as 100 million either way.

Country	Peak Year	Births then	Births 1969 (or latest available year)	Percentage decrease
Asia:				
Ceylon	1962	371,000	270,000 (1968)	27·2
Hong Kong	1962	119,000	83,000	30·2
Jordan	1966	94,000	69,000 (1968)	26·6
Pakistan *	1963	5,103,000	4,950,000 (1965)	3·0
Singapore	1959	64,000	45,000	29·7
Taiwan	1963	424,000	353,000	16·7
Africa:				
Algeria	1966	562,000	530,000 (1968)	5·7
Mauritius	1963	28,000	22,000	21·4
U.A.R.	1966	1,235,000	1,195,000	3·2
The Americas:				
Barbados	1960	8,000	5,000 (1968)	37·5
Canada	1959	479,000	371,000	22·5
Cuba	1964	264,000	232,000 (1967)	12·1
Jamaica	1966	71,000	65,000 (1968)	8·4
Puerto Rico	1962	80,000	68,000	15·0
Trinidad & Tobago	1962	34,000	28,000	17·6
U.S.A.	1961	4,268,000	3,571,000	16·3
Chile	1963	292,000	283,000 (1967)	3·1

trol') until their birth-rates are seen to be dropping. This is like saying to a patient, 'How dare you become ill! You shall have no medicine until you get better.'

If any rich nation adopted the Ehrlich-Paddock policies, which happily they show no signs of doing, the result would surely be the very explosion of population which these men fear. Birth-rates are kept down, not only by medicines and birth-control devices, but also by improved agriculture, clean streets, good schools and efficient hospitals. Fear and insecurity mean a high birth-rate. Confidence in the future means a low one. Developing nations cry out for the social and technical innovations which will bring these benefits—and they are steadily getting them. U.N. projections suggest that the world's 2 per cent annual population growth-rate will begin to slacken in about 1980; it is expected to stay at between 1·5 and 1·9 per cent for a period of ten to fifteen years after

* Both East and West Pakistan, including what is now Bangladesh.

that. Then it will slacken rapidly. Most demographers are convinced that during the next century, population will stabilize at about 10 billion. The figures of declining birth-rate which tell us that Malthus was accurate in predicting the effects of 'moral restraint' remove the last of the somewhat irrational doubts which have been expressed about the prospects for intelligent life on this planet. There is no limit to growth, and there is no limit to what the developed nations can accomplish.

3

Three Assumptions

The very phrase 'developed nations' sounds somewhat arrogant, since it implies that the world's most advanced countries have reached the end of their possible development, and that they have nothing more to learn or achieve. This is very far from being true. People in all ages are inclined to imagine that they have reached the peak of human achievement. Their descendants prove by their own accomplishments that their predecessors were wrong—but then they themselves often fall into the same complacent error. We ourselves are certainly guilty of it. A mere century and a half has elapsed since the Industrial Revolution, and we assume that we have now reached the height of our technological power. Our descendants will laugh at this belief, just as we laugh at the Victorians for believing exactly the same thing.

Is there, in fact, any limit to the possibilities of technology? We cannot answer this question with finality—nobody, in any age, will be able to—but it might be instructive to see how people in the past viewed their technical future with a conservatism perhaps even greater than our own. From the earliest times, nearly every advance in technology has been made against protests that it was impossible, invariably from people claiming expert knowledge, and almost invariably they were groundless. Arthur C. Clarke, possibly the most far-sighted philosopher of science of our age, declares as part of his 'First Law': 'When an elderly and distinguished scientist tells you that something is impossible, he is almost certainly wrong.' 'The expert can spot all the difficulties,' Clarke goes on to explain, 'but lacks the imagination or vision to see how they may be overcome. The layman's ignorant optimism turns out, in the long run—and often in the short run—to be nearer the truth.'[1] I have compiled a list of some of the more absurd of these historical negative predictions.[2]

Napoleon, preparing to invade Britain from his great camp at Boulogne, was approached by a down-at-heel American engineer named Robert Fulton, who explained how the blockading British fleet could be defeated. 'What, sir,' the Emperor is reported to have snapped after listening to him impatiently for a few minutes, 'you would make a ship sail against the wind and currents by lighting a bonfire under her decks? I pray you excuse me. I have no time to listen to such nonsense.' Had he listened, he might well have conquered Britain. The first primitive steamships, much as Fulton had foreseen them, appeared soon after Napoleon's death. But the experts saw little future in them. 'Men might as well expect to walk on the Moon as cross the Atlantic in one of those steamships,' stated the eminent Professor Dionysius Lardner.

Military men have always been quick to declare that things are impossible. One reason why the Roman Empire decayed may have been the refusal of its leaders to interest themselves in science. Their attitude is typified in a statement by Julius Frontinus, Rome's leading military engineer in the time of Vespasian. 'I will ignore', he wrote, 'all ideas for new works and engines of war, the invention of which has reached its limits and for whose improvement I see no further hope.' Soldiers in all ages have tended to see 'no hope for improvement'. It is a cliché that they are always well prepared for the previous war. Hundreds of thousands of men died from the combined effects of machine guns and barbed wire during World War I—because nobody had studied the Russo-Japanese War of 1904–5, in which exactly the same thing happened on a smaller scale. In our own time, President Eisenhower, himself a general, could not understand that something important had happened when the Russians launched the world's first spaceship, Sputnik 1, in 1957. 'The Russians have put a small ball up in the air,' he told a press conference. 'That does not raise my apprehensions one iota.'

Civilians can be just as negative. The introduction of railways led to arguments just as fierce as those that have accompanied the arrival of supersonic aircraft. A remarkable letter of complaint has survived from Governor Martin van Buren of New York to President Andrew Jackson in 1829:

As you well know, Mr President, 'railroad' carriages are pulled at the enormous speed of 15 m.p.h., by 'engines' which in addition to endangering life and limb of passengers, roar and snort their way through the countryside, setting fire to the crops, scaring the livestock and frightening women and children. The Almighty never intended that people should travel at such breakneck speed.

This kind of alarm has long been traditional in matters of transport. The astronomer Ptolemy declared in the second century A.D. that no man could cross the equator, since the Sun's vertical rays would boil the oceans and set wooden ships alight. It took Europeans twelve centuries, and the daring voyages of Prince Henry the Navigator of Portugal, to prove that Ptolemy was a humbug. Aeroplanes were long regarded as impossible. They involved 'flight by a heavier than air machine', the ultimate impossibility. As the astronomer Simon Newcomb wrote in 1903, 'aerial flight is one of that class of problems with which man will never be able to cope.'

A few months later, Orville Wright flew the first powered aircraft at Kitty Hawk, North Carolina. There was no publicity at first. Newspaper editors refused to print what one of them called 'this ridiculous story'. A few weeks later, when it was established that Orville really had flown, one thing was authoritatively declared: no plane could ever take the weight of a passenger. So Orville took along his brother Wilbur as a passenger on the next flight.

Well, so men could fly for short distance; but it would be no use to anybody. The engineer Octave Chanute wrote a famous article about the two-man flight of Orville and Wilbur Wright. Copies of the issue of *Popular Science Monthly* which carried it are now a valued collector's item. 'This machine may even carry mail in special cases,' he wrote. 'But the useful loads will be very small. The machines will eventually be fast, they will be used in sport, but they are not to be thought of as commercial carriers.' Eleven years later, in 1914, the first air passenger service was opened between two towns in Florida. All right, then, so the thing can carry passengers. But let's keep a sense of proportion! Shortly before World War I, the astronomer, William H. Pickering,*

---

* In the cause of fair play for Pickering it should be said that this person is no relation to William H. Pickering, now director of Caltech's Jet Propulsion Laboratory at Pasadena, nor to James Pickering, author of the indispensable book, *1001 Questions Answered about Astronomy*.

went into battle against some foolish ideas. 'The popular mind', he wrote, 'often pictures gigantic flying machines speeding across the Atlantic carrying innumerable passengers. It seems safe to say that such ideas must be wholly visionary. Even if a machine could get across with one or two passengers, it would be prohibitive to any but the capitalist who could own his own yacht.' Today, millions of passengers cross the Atlantic each year at 600 m.p.h., a speed which will soon be more than doubled by supersonic jets. The critics later proclaimed an absolute limit for all aircraft—660 m.p.h., the speed of sound at high altitudes. Learned professors wrote equations proving that it was impossible to exceed this speed. Disobligingly, in 1947, Captain Charles 'Chuck' Yeager of the U.S. Air Force flew his rocket plane *Glamorous Glennis* at 670 m.p.h.

Plans for space travel attracted similar derision. In the 1920s the citizens of Worcester, Massachusetts, became alarmed at the noisy experiments of Robert Goddard, who was trying to launch small rockets. There were frequent complaints to the police, and threats to tar and feather him, after his craft had exploded or fallen from a height on to neighbouring property. Yet America owes its successes in space more to Goddard, the inventor of liquid-fuelled rockets, than to any other individual. His suggestion that rockets could one day fly through the vacuum of space and reach the Moon earned him an attack from the *New York Times*. 'He seems to lack the knowledge [about vacuum] ladled out daily in the high schools,' said a contemptuous editorial.[3] The article went on, with surprising ignorance of Newton's Third Law, to declare that Goddard's proposed Moon rocket would 'need something better than a vacuum against which to react'.* Forty-nine years later, on the morning in 1969 when Neil Armstrong and his colleagues left Cape Kennedy for the Moon, the *New York Times* very honourably printed a formal apology to the long-dead Robert Goddard. In Britain, in the 'thirties, members of the newly-formed British Interplanetary Society did much of the theoretical groundwork necessary for a flight to the Moon.

* Newton's Third Law, which governs the behaviour of rockets and jet engines, states that 'for every action there is an equal and opposite reaction.' In other words, a rocket engine needs no air against which to push itself. The blast of gas pouring from its nozzles itself propels the vehicle.

They earned nothing but abuse from the scientific establishment. 'I was amazed at the half-baked logic that was used to attack the idea of space flight,' one of them recalled. 'Even scientists who should have known better employed completely fallacious arguments to dispose of us. They were so certain that we were talking nonsense that they couldn't be bothered to waste sound criticism on our ideas.'[4] One interesting comment in this vein has survived. The British scientist Professor A. W. Bickerton declared in 1926:

> This foolish idea of shooting at the Moon is an example of the absurd lengths to which vicious specialisation will carry scientists working in thought-tight compartments. To escape the Earth's gravitation a projectile needs a velocity of 7 miles per second. The thermal energy at this speed is 15,180 calories. Hence the proposition appears to be basically impossible.

Endless examples could be found of scientific predictions made in the last 100 years that were even more foolish than this. In 1899 the director of the U.S. Patent Office urged President McKinley to abolish the Patent Office along with his own job because 'everything that can be invented has been invented'. Lord Rutherford, having helped to split the atom in 1932, announced that he could see no practical use for his discovery. Critics laughed at Marconi when he proposed that radio messages could cross the Atlantic. It was supposed that to do this he would need a radio reflector as large as the North American continent. Soon after Alexander Graham Bell demonstrated the use of the telephone in 1876, Sir William Preece, chief engineer of the British Post Office, was asked for his comments. He made this remarkable reply: 'No, sir. The Americans have need of the telephone—but we do not. We have plenty of messenger boys.' Lord Kelvin, the great Victorian physicist, dismissed with contempt the now accepted theory of evolution: 'We find something at every turn to show the utter futility of Darwin's philosophy.' *

I have not listed all these anecdotes simply to poke ridicule at men, of whom many, apart from these aberrations, have con-

---

* The two men strongly disliked each other. Darwin used to refer to Kelvin as 'that odious spectre'. Their dispute on evolution, which hung on the age of the Earth, was later settled in Darwin's favour by the discovery of radioactivity in rocks. It is hard to comprehend Kelvin's opposition to evolution. Did he suppose that man arrived on this planet by magic?

tributed to large parts of our knowledge and civilization. But the very falsity of their negative predictions confirms Clarke's law and hints at the direction in which man is heading. Sir George Edwards, in his 1958 presidential address to the Royal Aeronautical Society, suggested that man's progress could be usefully measured by an index which has been called the 'speed exponential'. According to this system, the rise of human technology is measured by the distance which a person could normally travel in one 12-hour day.[5] For hundreds of thousands of years, this represented a pedestrian, who could do about 20 miles in a day. (This is an average, not a record. British infantry, before World War I, sometimes marched between 30 and 35 miles in a day, but their 'day' was often as much as 15 hours). The taming of the horse, a few thousand years B.C., doubled the average 12-hour distance to 40 miles. The road improvement of the Industrial Revolution brought also the stage-coach system, in which fresh horses could be acquired along the route. This innovation raised the 12-hour average to 75 miles. Railways later increased it still further to 550 miles. Aircraft made a still more radical improvement. By 1940, no less than 1,250 miles could be covered in a day, slightly less than the distance from London to Athens or from New York to Houston. The next two decades and the jet engine quadrupled this figure to 5,500 miles, so that within a single day by 1960 we could travel between London and Johannesburg or between New York and Beirut. Supersonic craft in the mid-'seventies will carry us more than 10,000 miles in a day, from London to Canberra or from New York to Sydney. This is a very great speed compared with the stage coach or the railway, but it is nothing to what will be achieved by residents of an orbital space station. These people, in a low orbit of 300 miles, will cover 240,000 miles in a 12-hour day, during which they will see nine sunsets. A spaceship to Mars will have a wearisome journey of many months unless it doubles this speed to a maximum of at least 50,000 m.p.h., enabling the astronauts to cover 600,000 miles in 12 hours. This is no fantasy. Manned expeditions to Mars seem almost certain to take place during this century, possibly before 1990. America's decision, taken early in 1971, to build the re-usable Earth-to-orbit shuttle system (thus cutting launching costs by some 80 per cent), and her agreement with Russia to carry out joint exercises in Earth orbit, make it increasingly probable that the two nations will ultimately send joint manned missions to the interior planets. And yet I predict that all these vast expenditures will compare in scale with the colossal space activities of the next millennium in a similar ratio to the fitting out of a stage coach against the 25-billion-dollar Apollo programme.

How can we ever afford to do such things? The gross national product of America, the richest nation on Earth, was a trillion dollars in the fiscal year 1970–71,* and we are suggesting that in a mere few hundred years she will cheerfully embark on space projects that will, by themselves, cost the 1971 G.N.P. in its entirety. Only a nation of madmen would consider spending their whole G.N.P. for a single year, or any large part of it, in a single enterprise of uncertain value. But the G.N.P. is not a constant. The American economy, at the time of writing, is growing at about 4 per cent a year. Assume that the long-term average rate will be about 3 per cent. A glance at compound interest tables tells us that anything that grows at 3 per cent a year will double every 23 years. The effect could be something like this:

| | U.S. G.N.P. |
Year	(Billions of dollars)
1971	1,000
1980	1,300
1990	1,750
2000	2,360
2050	10,330
2100	45,245
2150	198,173
2200	867,998

This is not nearly so fantastic as the extrapolations of the environmentalists which warn of a world population of 126 trillion in the year 2500. The principle of some kind of exponential growth in national or global wealth is valid, whereas a similar growth rate in population is not, for reasons explained above. Leaving aside the forms which these assets will take, it is evident that many nations, in the coming centuries, are going to become immensely

* I.e. 'real' gross national product, which allows for inflation.

rich as their technology grows. We need only look back a few centuries to see the great progress that has been made. Suppose that in the time of King George III we had known how to make aeroplanes that could fly. Some enthusiast might have proposed the building of a fleet of jumbo jets; after all, they might have been used for carrying troops and artillery to the American war. But the British government would have rejected this plan for a single reason; the fleet would have cost nine or ten times the country's gross national product. Today, such a fleet has been built costing a tiny fraction of a national G.N.P. The power to initiate great enterprises grows with wealth.

Freeman J. Dyson, of the Princeton Institute for Advanced Study (of whom there will be much later), made this point vividly in a 1968 lecture on the possibility of building spaceships for flight beyond the Solar System to the stars. To develop and build a manned starship, he admitted, would probably cost 100 billion 1968 dollars:

> Nobody in his right mind would consider building such ships at a time when our gross national product is only a few times the cost of one of them. But if we are thinking on a time scale of centuries, our G.N.P. is far from being a fixed quantity. If the human race continues its economic growth rate of about 4 per cent a year, we shall have a G.N.P. a thousand times its present size in about 200 years. When the G.N.P. is multiplied by a thousand, the building of a ship for 100 billion dollars will seem like building a ship for 100 million dollars today. We are now building a fleet of Saturns which cost about 100 million dollars each. It may be foolish, but we are doing it anyhow. On this basis, I predict that about 200 years from now, barring a catastrophe, the first interstellar voyages will begin.[*6]

This last statement might be re-written: if there is a catastrophe, interstellar voyages will probably be delayed until about 400 years from now. For, as we have seen, the time-scales of the Hertzsprung-Russell Diagram allow for almost limitless numbers of

* In a sense, Dyson was over-pessimistic by 197 years. Our first starship was launched in 1971. It was Pioneer 10, a U.S. unmanned craft which, after passing Jupiter, was due to leave the Solar System. It bore a coded plaque to inform any aliens of our existence. It has been followed by another such craft, Pioneer 11.

catastrophes. For convenience, I have called this book *The Next Ten Thousand Years*. I say that such-and-such an event will take place around 2100, and another by 2500, and so on. But these dates are not important. For although catastrophes may prevent the events predicted from happening during the next 10 millennia, if we accept three assumptions, *there is a probability of much greater than 99 per cent that they will happen sooner or later*. In a period of 6 billion years, almost anything that can happen will happen. And yet I make three assumptions. Three different kinds of disasters could bring all humanity's daring aspirations to a final end. The reader can assess the chances for himself that my assumptions will prove valid:

1. That there will be no substantial change in the Sun's radiation.

2. That the Solar System will not be invaded by a hostile or a meddlesome alien technology.

3. That there will be no fundamental change in human nature —or, more precisely, that the human reaction to stimuli will remain constant.

Assumption No. 1 worries me the most. It seems the most likely, or rather the least unlikely, to break down. The Diagram, although much refined since it was first conceived in 1913, is still something of a rough and ready document. The chances that it could be wrong in the case of any single star, like the Sun, are very remote, perhaps one in many millions, but they are not zero. A more serious danger is malicious interference. Blowing up the Sun, and destroying all life on Earth in a single action, would be a perfect *götterdämmerung* for a besieged warlord. It would be a spectacle to surpass all others; but the fact that nobody would survive to witness the effects might, paradoxically, make the action even more attractive to warped or highly 'poetic' minds. One can easily cite examples of this kind of pyromania. Hitler is said to have exclaimed several times while taking refuge in his bunker in 1945, 'I wish there was a bomb that could blow up the whole world!' Tiberius Caesar used to quote with ominous relish the line from Euripides: 'When I am dead let fire the earth consume.'

At the court of Nero, three imperial reigns later, somebody again quoted this line. Nero insisted that its first part should read:

'While I yet live'. He soon converted fancy into fact, and ordered Rome to be set on fire. According to Suetonius, he watched the fire from a tower, enraptured by what he called 'the beauty of the flames'. He then put on his tragedian's costume and sang *The Fall of Ilium* from beginning to end. His conduct, if classical writers are to be believed,[7] shows a state of mind that would delight as much in a cosmic pyre as in the destruction of a city. The astronomers Carl Sagan and Iosif Schlovskii suggest that a powerful civilization of the distant future might be able to blow up stars by means of a super-advanced laser with a power output of 10 trillion kilowatts,*[8] If each square centimetre of a small area of the Sun's surface could be bombarded by 10 billion ergs of gamma radiation per second, a chain reaction might result, convulsing the whole Sun into a supernova explosion that would obliterate the Earth. It is also conceivable that such an action might one day be threatened by blackmailers, hijackers on a global scale, or that the detonation of an enemy's sun could become standard military procedure in an interstellar war. It must be hoped, however, that the leaders of rival stellar systems would refrain from this sort of attack from the same fears of retaliation that restrain nations today from nuclear war. The construction of a Sun-destroying laser gun would be very costly in any age and difficult to keep secret. The chances that Assumption No. 1 could break down through human malice must still be considered fairly slim.

We have no data with which to evaluate Assumption No. 2. Several writers have tried to show that the Earth was visited by aliens in the remote past, but we lack hard evidence. As for the present, I am always ready to be convinced of the existence of flying saucers—if only one of them would land, and its inmates would pose for photographs, perhaps demand an interview with the Prime Minister, or at least hold a press conference. It is difficult in any case to see *why* a species should send an expedition tens of trillions of miles through space, with no other purpose

---

* Sagan and Schlovskii do not suggest this as a warlike scheme. They propose instead that artificially-created supernova explosions could enable stars to be mined for their heavy elements (see (Chapter 13). Ten trillion kilowatts is a tremendously large power output. It is about 1,000 times the present power consumption (per second) of modern global civilization.

than to alarm airline pilots. The chances of a breakdown in the second assumption must, in the absence of any data, remain very low.

The third assumption seems even stronger than the other two. We cannot summarize precisely what human nature is, but it appears certain that it has not changed in any fundamental way since the earliest recorded epochs if human history. Egyptians in the time of the Pharaohs had much less information than we have—just as we know much less than will our descendants—but the general laws of psychology, the way in which a person will react to a stimulus in a given situation, have been the same since the dawn of *homo sapiens*.

This generalization might seem to beg several questions. What *are* these general laws of psychology of which we talk so glibly? How will a person react to a certain situation in given circumstances? We still have only the vaguest notions. Despite huge numbers of books and research papers on the psychology of mobs and of individuals, we still have no general picture of 'man'. Hereditary factors, it now appears, have a much stronger influence on our behaviour and intelligence than the environments in which we are brought up.[9] In addition to infuriating many social thinkers, this scientific conclusion has made analyses of human nature very difficult and complicated. People differ in their behaviour and thinking far more than do the members of any other known species. Some human reactions can be predicted with 99 per cent certainty. If an employer slaps the face of his shop steward there will be a strike. If one country invades another there will be a war. But any reasonably intelligent Pharaonic Egyptian could have told us simple things like these. Of more complex cause-and-effect chains of events we are progressively more ignorant. Human nature remains constant, but any general characterization of man can only be made by compiling a vague string of adjectives. Let us try such a characterization. A historian, much given to purple prose, once wrote the following description of King Henry VIII:

This monarch was sincere, open, gallant, ambitious, wise, zealous, liberal, intrepid, inflexible and courageous. But with these virtues, he combined the vices of violence, cruelty, profusion, rapacity, injustice, obstinacy, arrogance, bigotry, presumption and caprice.[10]

Leaving aside the question of whether this is a fair description of Henry VIII, it does appear to be a roughly accurate representation if applied to the human race in general. No doubt numerous other qualities should be added, but many will be mere subdivisions of these. Our prospects for racial survival may indeed look more promising if we can say that for every decision-maker who is 'arrogant, bigoted and presumptuous', there is another who is 'wise, zealous, and intrepid'.

So much for human nature, little understood but unchanging, whose 'ambition, wisdom, and zeal' have raised us to unparalleled heights of economic and technical power. Some alarming stories have appeared. It has been suggested that biological controllers, with the noble aim of suppressing crime, envy and other of Henry VIII's alleged vices, may in the future establish some kind of gene technology, so that hereditary influences, and through them all behaviour, could be controlled. Indeed, if this ever happened, it could mean the collapse of my third assumption and the impossibility of any predictions. A race whose qualities were controlled by some bureaucratic agency, instead of by the random mixing of genes that we have known throughout history, would have little interest in scientific speculation or of going to the stars.* Happily, as Gerald Leach argues in his book *The Biocrats*, such a policy appears extremely unlikely to be undertaken. Such a programme, by suppressing individualism and ingenuity, would guarantee incompetence. It would mean also a total and permanent loss of freedom. For this reason, it would be intolerable. As Leach explains:

> For 5,000 years it has been perfectly possible for power-mad rulers to breed selectively and cull human beings into specialized types, just as humans have bred cows and dogs. None have done so because none of their subjects would stand for it for a moment and, anyway, if one wants to make specialized men it would be far easier and quicker to train them.[11]

* James D. Watson, co-discoverer of the genetic code, has given many lectures warning scientists and governments against the social dangers of any programme of gene control.

The Nobel-laureate biologist Sir Peter Medawar puts it even more emphatically: 'The manufacture of super-men by cross-breeding is unacceptable today, and the idea that it might one day become acceptable is unacceptable also.'[12] Lest it should be feared that these specialists might be proved just as wrong in their predictions as the other people I have quoted, we may reflect that, judging by experience with animals, to change a species by breeding would take a minimum of 20 generations, or about 600 years. It is doubtful whether any dictator or zealous social reformer would have any interest in waiting that long.

What, then, can be expected to happen if my three assumptions hold true? Contrary to the Club of Rome's belief, there are no 'limits to growth'. There is no reason why our global wealth, or at least the wealth of the industrial nations, should not continue to grow at its present annual average of 3 to 5 per cent indefinitely.[13] Even if the Earth's resources prove ultimately to be finite, those of the Solar System and of the great Galaxy beyond are, for all practical purposes, infinite. The argument which has followed the Club of Rome's report may be seen as the first of many 'industrial crises'. These 'crises' will take the form of renewed questioning about the desirability of economic growth. But politicians, in nearly all cases, while vaguely accepting the idea that there must ultimately be a limit to growth, will insist that such limits will only be reached in the far distant future. And when that distant future arrives, we can be confident that politicians of the day will teach exactly the same conclusions. They will act in the same way as have the industrial nations who in our time have rejected the implications of the Club's report and resolved to continue to boost their economies at the fastest possible rate.*

Not choosing to stagnate, man will carry his economic activities into space. When predicting that the next century will see a colossal increase in manned space activity, I will try to avoid what

---

* Anyone who doubts this should study the speeches of almost any Finance Minister, and the speeches at any conference of the International Monetary Fund. Even Communist China, which long preferred a low-level agricultural technology, has since 1971 been raising this level with massive purchases of civil jet aircraft, and with the development of many other advanced projects.

Freeman Dyson calls the 'Gee Whiz' approach to space, which simply involves extrapolating the technical achievements of the last 10 years into the next century, and suggesting exotic places where the American flag might then be flying. ('We'll have men on Mars by 19–. Gee whiz!') Let us return to N. S. Kardashev's 'phases' of the growth of cosmic civilization, referred to briefly in Chapter 1. Kardashev sees our present way of life as a 'Phase 1 civilization', in which we exploit the resources of a single planet. We shall enter Phase 2 when we have mastered the resources of our entire Solar System, using our industrial technology to dismantle whole planets and arrange the fragments at more convenient orbits round the Sun. In this way, by gradual steps, we shall come to harness the life-giving power of the Sun's radiation, of which the Earth at present only receives one part in 500 million.* A civilization which achieved such things would have to be about 10 billion times wealthier and hence more powerful than is ours today. Such a huge increase in wealth might seem a ridiculous proposition, but I have tried to show that it is not. Some economists believe that we may achieve a long-term average annual growth-rate of 5 per cent. This seems extremely optimistic. Let us assume a more modest 3 per cent. A growth-rate of 3 per cent means a doubling every 23 years. If this 3 per cent rate was maintained, on average our civilization would then be about 10 billion times richer every 730 years. Although frequent recessions and periods of stagnation will no doubt prevent this level of wealth from being reached within much less than 1,000 years, man is inevitably going to become hundreds of thousands of times richer within the next few centuries than he is now, and that factor of 10 billion will eventually be attained.†

The transition from Phase 1 to Phase 2 has been in progress

* These vast schemes are described in Chapters 12 and 13.

† For a typical environmentalist misunderstanding of this position, see John Loraine, *The Death of Tomorrow* (Heinemann, London, 1972), p. 265: 'Two hundred years from now, the American G.N.P. will be 2,500 trillion dollars, or some 2,500 times the present G.N.P. The situation will indeed be glorious, for every citizen of that country will then be a millionaire. But the glory will never be attained, for long before that time the Earth will have been denuded, its sources of energy will have vanished and its environment will lie in ruins.' If the Earth was alone in the Universe and its resources were unique, Dr Loraine's view would be fair. But neither of these things is true.

since the first satellite was launched in 1957. Sputnik 1 has been followed by many hundreds more satellites, of increasing size and complexity. Men have flown in innumerable orbits round the Earth, a large number of craft have visited neighbouring planets, one unmanned ship has headed out of the Solar System towards the stars, 12 men have walked on the Moon, and 9 more have done experiments in new industrial processes in the weightlessness of the Skylab orbital factory. Perhaps most significant of all, the agreement by America and Russia to cooperate in space may prove the most important and beneficial check to the belligerence of the Cold War since it began.

Irrespective of any short-term cut-backs in space budgets, it is certain that this trend of increasing exploration and exploitation of space will continue exponentially for the next millennium. On occasions, men may be elected to high office with a determination to shut down space activity completely.* They may succeed in doing so for periods of 10 years or even longer. But the demands of economics and technology will exert an ever stronger pressure for its resumption. Of all celestial bodies accessible to us, the Moon is the most easily reached and exploited. The Apollo landing missions have ended, but within two decades or so men will return to the Moon in far more elaborate expeditions. An interesting view on this point has been expressed by Lord Shackleton, president of the Royal Geographical Society, who has an extra distinction in being the son of Sir Ernest Shackleton, the leader of two desperate expeditions to the Antarctic. Shackleton junior is himself a noted explorer as well as a respected scholar, having participated in an expedition to the icebound Ellesmere Island in northern Canada. He was interviewed on the day that the last Apollo mission returned to Earth in 1972:

If the history of exploration shows anything, it is that once man has discovered any new territory, he will return to it, again and again, finally to set up a permanent base there. I am sure that

---

* It is interesting that President Eisenhower, before 1957, was hostile to *any* space activity. Senator George McGovern, in his 1972 Presidential campaign, promised to stop all future *manned* space missions. The success of unmanned satellites in saving lives and improving communications had by this time became obvious. The growth of technology often thwarts reactionary policies.

the same thing will happen with the Moon in a few decades. Man will go back to the Moon, simply because it is there, because the challenge of exploration is so enormous, and because it is of immense scientific interest. Man is returning again and again to Mount Everest, since it was first climbed in 1953, and the scientific results from it are very much less than from the Moon.[14]

Shackleton's view is almost certainly prophetic. Not only does this nearest world to ours provide a perfect laboratory for studying the Universe; it also represents a ready-made space station with a surface area as large as Africa. Let us examine its probable future uses.

4

The Beckoning Moon

The Moon has for thousands of years had a powerful effect on human imagination. Man has always tended to imagine it somehow as a much *better* place than Earth, and nearly every ancient civilization had its Moon deity.* No doubt its silvery-golden colour, as it scudded above the murky cloud-banks of a troubled world, did much to encourage the impression that the Moon was 'pure' while the Earth was 'foul'. This idea is vividly expressed in the epic poem *Orlando Furioso*, completed in 1532 by the great satirist of the Renaissance, Ludovico Ariosto. His hero Astolpho is disgusted by life on Earth, where he finds that time and wealth are misspent, vows are broken, prayers are unanswered, tears are fruitless, ambitions are frustrated, and desires are unfulfilled. But Astolpho solves these problems by flying to the Moon. There, he finds that bribes are hung on golden hooks, princes' favours are kept in bellows, and wasted talent is conveniently preserved in vases.[1] Alexander Pope enlarged on this theme in 1714 in *The Rape of the Lock*. When the Lock disappeared, according to Pope:

> Some thought it mounted to the Lunar sphere,
> Since all things lost on Earth are treasured there;
> There heroes' wits are kept in pond'rous vases,
> And beaux' in snuff-boxes and tweezer-cases.
> There broken vows and death-bed alms are found,
> And lovers' hearts with ends of ribbon bound,
> The courtiers' promises, and sick men's prayers,
> The smiles of harlots, and the tears of heirs,
> Cages for gnats, and chains to yoke a flea,
> Fried butterflies and tomes of casuistry.[2]

* As, for example, Ashtoreth in Carthage and Canaan, Astarte in Egypt, Artemis in Greece and Diana in Rome.

These are but two examples of how poets throughout the ages have imagined the Moon as a sort of mystical answer to all human problems, a golden land where all wrongs are righted. People have always felt instinctively that the Moon was part of man's property. Oscar Wilde shows this with an amusing anecdote. While travelling to Louisiana after the Civil War, he found a gloomy belief that the war had devalued everything. On remarking to his host, 'How beautiful the Moon is tonight,' he received the reply, 'Yes, but you should have seen it before the war.'

It has only become realized in quite modern times, however, that we are extremely fortunate to have such a large Moon at all. Although, with its diameter of 2,160 miles, ours is only the fifth largest moon in the Solar System, being surpassed in size by Jupiter's Ganymede and Callisto, by Saturn's Titan and by Neptune's Triton, it is much larger in proportion to its parent planet than any of these. Ganymede and Callisto have, respectively one twelve-thousandth and one twenty-thousandth of Jupiter's mass. Saturn is 4,000 times more massive than Titan, and Neptune has 750 times Triton's mass. Mars is many millions of times more massive than either of her two little moons. But our Moon is huge in proportion to Earth. It has no less than one eightieth of the Earth's mass.

Many astronomers in fact deny that the Moon is a 'moon' at all. They say instead that the Earth and the Moon are a 'double planet'; that is, two planets which happen to be circling one another instead of moving in separate orbits round the Sun. Others dismiss this argument as meaningless; they believe that asking whether the Moon is a planet or satellite is like a waitress asking a customer if he wants his toast under his eggs or his eggs on top of his toast. Whatever the truth of this matter, it must be seen as a curious coincidence that the one planet in the Solar System that is rich with life should have very much more than its fair share of moon. Our large Moon has always been of benefit to us. It cleans the oceans daily with its tides, and it has enriched the culture and the poetry of many races; it has also been used through the ages for practical purposes. As early as 8,500 years ago, men carved notches on animal bones to mark the phases of the Moon in order to calculate the lengths of the seasons more precisely.[3] The Sioux Indians painted their buffalo skins with notations indicating Moon-phases to subdivide the seasons. Their winters would include such dismal periods as 'Snow Moon' and 'Hunger Moon'. Their springs were a more cheerful time with 'Awakening Moon' and 'Salmon Moon'. Autumn included 'Leaf-Falling Moon', and 'Bison-Fighting Moon'. Then it was back to winter again with 'Long Night Moon'.[4]

More lately, the Moon's existence has helped give birth to the idea of space travel, thus beginning a new and exponential increase in our technology. Now, it promises to carry that process incomparably further. The Moon will be more in the future than a mover of tides and a beacon for lovers. It will provide man with the land and the environment to achieve one of his greatest advances in technical and industrial growth.

The Moon certainly does not at first present an agreeable sight to visitors. It is a battered, forbidding place of mountains and deserts. Edwin 'Buzz' Aldrin, the second man to walk on its surface, remarked that he had never seen such 'magnificent desolation'. It took several months of experiments with the Moon-rocks returned from the first three manned landings of the Apollo programme before it was realized that the Moon might be much less inhospitable than it had seemed.

Men have long wanted to establish permanent or semi-permanent colonies on the Moon. But they have always been faced with the difficulty that the Moon appears to be devoid of water. Bringing water up from Earth would make such colonies prohibitively expensive. The early Moon landings seemed to confirm this gloomy impression. For all their fierce debates about the origin and history of the Moon, the quarrelsome Moon scientists tended to agree about one thing; the Moon's surface and outer crust was dry and barren. The first Moon-rocks, from Apollo 11 in 1969, were 'anhydrous'; they contained not a trace of water, or any signs to show that they had ever been in contact with water. As one geologist remarked after careful studies of the rocks returned by Apollo 11, the Moon appeared to be 'a million times drier than the Gobi desert'. It was obvious in retrospect that hydrogen was too light an element to be retained by the weak, one-sixth Lunar gravity, which would have permitted its escape into space soon after the Moon was formed.[5]

This is not to say that water does not exist inside the Moon. There may well be layers of ice beneath the surface, or even on the floor of deep craters where the Sun's rays never penetrate. An experiment of Apollo 14 in 1971 strongly suggested that geysers of water vapour sometimes bubble up from fissures in the surface. The astronauts of this mission to the volcanic Fra Mauro region left behind on the Moon's surface the first Suprathermal Ion Detector Experiment * (SIDE), whose function was to record and identify any escaping gases. At the third Moon science conference at Houston, Texas, in October 1971, two physicists, John Freeman and H. Kent Hills announced the results of SIDE, and the many theorists who had always believed in a dead, cold Moon were suddenly in a minority. Freeman announced that SIDE's radio signals had recorded a geyser-like outburst of water vapour that lasted twelve hours.[6] He and Hills were convinced that the eruption was of natural origin, and had nothing to do with any leaking from the descent stage of the astronauts' Lunar module which remained on the surface. 'There is a good possibility of liquid water in the Moon,' Freeman said. 'It may represent a potential benefit. We may be able to tap this source of heat energy and water.' But other scientists were sceptical about the experiment. They were unconvinced by Freeman's statement that the gas was hydrous.

But the would-be Lunar colonists are undismayed by any doubts about the existence of hydrogen. For liquid hydrogen, being roughly 100 times lighter per cubic inch than water, *can* economically be brought up from Earth. And having once arrived there, it can be recycled almost endlessly if combined with oxygen to produce water. And there is plenty of oxygen on the Moon. The Apollo rocks were rich with a double oxide of iron and titanium called ilmenite. A group of NASA scientists, discovering this, quickly sought a U.S. patent for a 'hydrogen reduction process'† which would provide Lunar colonists with unlimited quantities of water.[7] This process is brilliantly simple. A quantity

---

* Meaning simply a device for detecting and identifying hot gases.

† They are unlikely to get the patent, for geographical reasons. Any invention patented in America must have application within the United States or its territories, but the United Nations has ruled that no country has jurisdiction on the Moon or planets. Admiralty law precedents may be needed to evade this technicality.

of ilmenite is placed in a container suspended above the surface. Beneath it is a Solar reflecting mirror, which heats up the ilmenite to between 2,000 and 3,000 degrees Fahrenheit. Hydrogen gas is then introduced into the container through a tube. It reduces the oxygen atoms from the frying ilmenite into steam. The steam is exhaled from the ilmenite container into a cooling chamber, where it becomes ordinary water. This process can be operated on a huge scale on the Moon at no great cost, other than the expense of building the apparatus and importing the initial hydrogen. The NASA experimenters found that if Lunar soil containing ilmenite is placed at random in the container, 100 Earth-pounds of soil will yield nearly a pound of water. But a simple refinement would make the process much more efficient than this. If the container were given a strong magnetic field to separate the iron-bearing ilmenite from the rest of the soil, 100 pounds of ilmenite could yield no less than 14 pounds of water. Oxygen for breathing —each man will require three pounds per Earth day—can be extracted from the water. It will also be produced by agricultural crops under sealed, transparent domes.

Nor are these the only sources of oxygen on the Moon. Samples from the Apollo rocks show that oxygen is more abundant on the Moon than any other element. It has been estimated that 75 per cent of the Earth's and Moon's crusts alike, down to a depth of 10 miles, consist of compounds from only two elements, oxygen and silicon. Oxygen is found in the Lunar crust in the form of oxides. See page 62 for a comparison of the amounts of oxides in the Apollo 11 rocks with those found in terrestrial deep-sea basalts. [8]

A ton of oxygen can be extracted from 2·6 tons of iron ore, or from 1·1 tons of aluminium ore. This would provide 1,000 days' supply of oxygen for one man, without any recycling being necessary.

To what extent will the Lunar colony be self-sufficient? The hydrogen reduction process is unlikely to be 100 per cent efficient. This means that some liquid hydrogen will have to be brought up each year from Earth. This will not be prohibitively expensive, since, as I have explained, hydrogen is very light. Let us calculate the initial amount of hydrogen the colony will require. Assume that each person on the Moon needs 10 gallons of water per

Oxide	The Moon's crust %	Basalt on Earth %
Silicon dioxide	40·4	49·2
Aluminium trioxide	9·4	15·8
Iron oxide	19·3	8·2
Magnesium oxide	7·2	8·5
Calcium oxide	11·0	11·1
Sodium oxide	0·5	2·7
Potassium oxide	0·2	0·3
Manganese oxide	0·3	0·2
Titanium dioxide	10·9	1·4
Phosphorus pentoxide	0·1	0·1
All other oxides, including traces of water (hydrogen oxide)	0·7	2·5
	100·0	100·0

Earth-day.* The full payload of the Apollo lunar module is 15 tons. Fifteen tons of liquid hydrogen, assuming unlimited quantities of oxygen, will yield about 135 tons, or 30,240 gallons, of water. At a rate of 10 gallons per man per Earth-day, and assuming a colony of 1,000 people, the 15 tons of hydrogen would have to be recycled every three days, or, for 100 people, every 30 days. But far more water than this will be needed for industrial purposes, so we can multiply the initial hydrogen supply by 40—40 Apollo-size payloads, weighing a total of 600 Earth-tons, and yielding 1·2 million gallons of water. Assuming that the recycling plants do not break down, but allowing for a certain wastage of hydrogen, we can say that the colony will need 600 tons of liquid hydrogen initially, and perhaps 50 tons for every year thereafter, decreasing perhaps as recycling techniques improve. As for the other water component, oxygen, we have

* This estimate is a compromise between the 0·6 gallons which the average man drinks each day, and the 125 gallons which the average U.S. city dweller uses. But the latter figure includes his bathwater, the washing of his car, the watering of his garden, his cooking, and the industrial liquids involved in his job. Much of this water would be unnecessary or would be recyclable on the Moon.

seen that supplies on the Moon are almost unlimited. The Moon has enough ilmenite and other oxides to provide a colony of 10,000 people with drinking water and oxygen for tens of thousands of years.

We thus imagine a large colony of Lunarians with abundant drinking water and oxygen, but so far without electrical power to run their machines. But it now appears that this can be provided far more cheaply than hitherto imagined. For this purpose, there will be no need for nuclear reactors or Solar batteries or any of the other exotic kinds of Lunar power stations that have often been predicted. Electrical power may come from a system akin to old-fashioned steam generation, a simple idea worked out by Patrick Moore and Anthony Michaelis.[9] Four generating plants are placed in each of the four quadrants of the Moon's surface, in such a way that one of them always receives the full heat of the Lunar day. Each plant is connected by surface cable or by overhead pylon to all points on or beneath the surface where electric power is needed. At the plant, liquid nitrogen inside a tank is heated into energetic gas or 'steam' by sunlight, which on the airless Moon normally raises the temperature of surface rocks to more than 240 degrees Fahrenheit. (Liquid nitrogen is most suitable, since the Apollo rocks have shown that nitrates are abundant on the Moon. Liquid nitrogen has also the extremely low freezing point of minus 380 degrees Fahrenheit.) The nitrogen steam flows from the tank into a power-generating plant where it drives a turbine. It is then recycled into a second tank which is screened against the sun's heat. It is thus enabled to cool down again. It can then be transferred again to the first tank, where it can go through the whole process again, and so on indefinitely.

This scheme, like the hydrogen reduction process, has the virtue of extreme cheapness and simplicity. It requires no fuel, except for the initial supplies of liquid nitrogen, and no human assistance after it has been built except for repair and maintenance. Sunlight will activate the power plants automatically. Reflecting mirrors beneath the tanks might give an additional boost to the power by providing even greater heat. A switching system and the careful location of the four power stations in the Moon's quadrants will ensure that as soon as one power station no longer has enough sunlight for it to operate, its functions will be at once taken over by another. Moore and Michaelis at first imagined that two such power stations would provide the entire Moon with electricity at any one time, but they soon saw that however ingeniously the two power stations were located, at least 48 hours would elapse while one station was in darkness, and the Sun was only creeping over the other's horizon. At this point, neither would receive enough sunlight to generate power. This is a particular problem on a slow-rotating body like the Moon, where a day and a night each lasts 14 Earth-days. But when there are four properly placed steam-power stations, one of them will always receive the Sun's full heat.

The picturesque adventure of Astolpho, who finds that all things denied to man on Earth are granted to him on the Moon, may thus contain some truth. Intelligent use of the rigidly synchronous Lunar calendar, in which the Sun, blazing out of a cloudless sky, appears in full strength over any point at a precisely calculated second, will give men a virtually free source of electricity and water, all achieved by processes impossible inside the Earth's cloudy atmosphere. People are only slowly beginning to realize how great an asset the Moon is going to be. Back in 1961, soon after President Kennedy committed America to a manned Lunar landing, Arthur C. Clarke published a detailed magazine article suggesting that in the 21st century the Moon will be an asset 'more valuable than the wheatfields of Kansas or the oil wells of Oklahoma'.[10] An engineer working with NASA wrote to the magazine in question demanding to know whether the article was intended as a joke. Clarke replied indignantly to the engineer that there wasn't much hope for the space programme so long as *he* was involved with it.

As late as 1968, the astronomer Fritz Zwicky reported that plans to exploit the Moon still seemed fantastic to the man in the street and to the 'ordinary calcified scientist'.[11] In the coming decades, however, the Moon will seem increasingly attractive to the entrepreneur. Its much smaller mass gives it an environment quite different from ours. Its weak gravity will make possible the construction of buildings, magnetic railway tracks, spaceports and huge telescopes from materials so flimsy that on Earth they would crumble at the first breeze. Lunar telescopes, in particular, will enlarge our knowledge of the Universe to an extent that we now only dimly imagine. It is a reasonable expectation that within the first five years of Moon-based astronomy, we shall uncover more secrets in the heavens than we have done in the three and a half centuries which have elapsed since Galileo first peered at the sky through his primitive lens.

5

The Lunarians

Astronomy from telescopes based on Earth is approaching the end of its useful life. To an ever greater extent, the glare from millions of city lights and the electrical interference from hundreds of thousands of domestic machines are wrecking the delicate observations of the world's foremost optical and radio telescopes. Within another decade, the need to move telescopes into space or on to the Moon, away from terrestrial disruption, will become urgent. As urban sprawl closes in on lonely mountain observatories, it becomes increasingly difficult to take accurate photographs of faint star clusters and distant galaxies.[1] Mercury-vapour street lamps, for example, increasingly favoured by municipal authorities because they reduce the number of road accidents, are a particular nuisance to astronomers. They happen to be a strong source of ultra-violet radiation, which is in that part of the light spectrum which gives important clues to the composition of certain stars and galaxies. Hours of patient photographic work can be ruined if a city's lights and advertising billboards give off the same light characteristics as a star. When the roads of a large suburban area are paved with light-reflecting concrete, and tiny smog particles mirror the glare from huge cities, the increased brightness on the horizon of an observatory 50 miles away is like that of the rising moon. Floodlit buildings. intended to be visible for miles, and searchlights, stabbing into the sky at random, make the situation still worse. The 48-inch telescope at Mount Wilson, California, which before the war was famous for its views of far galaxies, has been forced to cease this kind of work by the dazzling glare from Los Angeles. Even the biggest optical telescope in the world, the 200-inch reflector at Mount Palomar, situated on a forested mountain in a remote part of California, may be seriously imperilled by 1980 by the brightening lights from Los Angeles and San Diego. Telescopes like Mount Palomar are a million times more sensitive to light glare than the unaided eye. The managers of Britain's huge radio telescope at Jodrell Bank are being forced to build a new type of receiver in a remote Welsh valley because Jodrell Bank itself is finding it difficult to distinguish its cosmic signals from those of the domestic machines of the thousands of new homes that now surround it.[2] Astronomers are seeking to build their telescopes at ever more remote locations. But as population increases and spreads, even this expedient will not be possible for long. If our promising exploration of the Universe is to continue, telescopes will have to be lifted clean away from terrestrial interference, into Earth-orbital stations and on to the Moon.

Even without interference from city lights and domestic electricity, views of the sky from inside the Earth's thick atmosphere are far inferior to those obtainable from space. It is astonishing that we have learned as much as we have about the Universe from inside our murky atmospheric ocean. Imagine the night sky as seen by a fish a few inches beneath the surface of the sea. The fish can see the full Moon clearly, but the Moon seems to dance around at each agitation of the water, changing its shape every second in wild convulsions. The stars are faintly visible in their correct places at those rare moments when the water is absolutely calm. But at all other times they appear to rush furiously across the sky in zig-zag paths and chaotic ellipses. The whole view is dimmed and obstructed by the water; however clear the night sky, the fish can only see it through a fog.

But above the surface, the scene leaps into clarity. The fog is gone, and celestial objects shine brightly at their correct distances from each other. Yet even on the stillest, most cloudless nights, the sky is not truly clear. A man outside the atmosphere in space can see thousands of times more stars than are ever visible from the Earth's surface. The improvement in visibility between a telescope at the surface and one in space is in fact as great as the difference between a telescope beneath the sea and one on the surface. The atmosphere screens, distorts and smothers the incoming signals from the stars. An entire portion of the light spectrum is blocked off from us. Ultra-violet radiation and X-rays are disrupted by ozone in the stratosphere and by oxygen at still higher levels. Heat shimmer in the air blurs the images of stars, making them appear to twinkle. The planets show indistinct edges and fuzzy surface features. There are rare moments when all air turbulence is stilled—just as the fish-astronomer enjoys rare moments of absolutely calm sea—and it is at these moments when most of the important astronomical discoveries have been made. But they seldom last long enough for the lengthy photographic time exposures that improve visibility a thousand-fold. Better results have been obtained from telescopes carried by balloons and sub-orbital rockets, but balloons cannot go higher than 20 miles, which is still partly inside the atmosphere, and rockets wobble and vibrate.

Several telescopes have been launched into Earth orbit, and have performed impressively. But, except for those aboard Skylab, they are operated from Earth by remote control, and are therefore of necessity limited. Orbital telescopes will perform far more efficiently when men are in space to manage them directly. Men will be needed to develop and inspect photographs and preserve the valuable photographic detail that is now lost when pictures are radioed down to the surface. Men can check exposure times, repair defective equipment, correct bad photographs by a second exposure, and perform many other chores that remote-controlled devices cannot. Much more sophisticated telescopes will be placed in orbit during the next two decades, as large manned space stations come into operation. But although attractive compared with surface observatories, they will have several serious disadvantages:[3]

1. All human movement on board, however slight, will be liable to disturb the space station enough to ruin time exposures, constantly necessitating corrections and re-starts.

2. The Earth will eclipse a large part of the sky, depending on the height of the space station's orbit. It will also create a glare in the sky comparable to that from Los Angeles which has ruined so many observations at Mount Wilson.

3. The intense radiation of the Van Allen belts surrounding the Earth will be liable to ruin many kinds of observations, and telescopes are likely to need several tons of shielding as protection from them.

4. Radio, television, and aircraft signals on Earth will disrupt radio telescope observations.

5. Rotation of the space station, which will be necessary to create gravity for the sake of the astronomers' health, will mean that stars move much too quickly across the sky for long time exposures. Counter-rotation of the telescope will be a crude solution, and much accuracy will be lost.

All these difficulties disappear if telescopes are placed on the Moon, preferably on the far side, so that the Lunar diameter of 2,200 miles can shield observatories from terrestrial radio noise. The expectation of Moon observatories being 'shielded by 2,000 miles of rock' has become almost a cliché among futuristic space writers. But the shielding will be of enormous importance, For 14 Earth-days, the period of the Lunar night, observatories will enjoy the most perfect conditions that are possible anywhere. An astronomer arriving for the first time to use a telescope on the Moon will immediately notice one beneficial difference from what he has been accustomed to at home. The stars will no longer appear magnified and distorted by their own brilliance. Paradoxically, the very brilliance of stars is a serious nuisance in terrestrial astronomy. The brighter the star, the more it assumes the shape of a disc when photographed through more powerful telescopes, an illusion caused by the diffraction of light rays as they pass through the atmosphere. It frustrates astronomers today to know that light messages from the stars have travelled towards Earth for trillions of miles through space—only to be almost ruined in the last hundred. In fact the stars are so distant from us that, however powerful the telescope, they still ought properly to appear as dimensionless dots. For the star-dots in an enlarged photograph of part of the Milky Way to become real discs representing the actual sizes of individual stars, the photograph would have to be still further enlarged until it covers the whole of Europe.[4] Only then could individual stars accurately appear as little pea-sized discs. Through a telescope on the airless Moon, none of this false magnification would occur.

Radio astronomy will have tremendous advantages on the Moon. The world's largest unified partially steerable radio telescope today is a huge construction, 1,000 feet in diameter, slung across an extinct volcano crater near Arecibo, in Puerto Rico. Its assembly platform is built to withstand tropical hurricanes and weighs 600 tons. No such structure will be necessary on the Moon, where there are neither wind nor storms. A radio telescope there can be built by simply stretching aluminized plastic across the floor of one of the giant craters. Cables for its focusing assembly can be suspended from the crater walls, and the entire construction can be as large as 50 miles in diameter.[5] For since everything on the Moon weighs only one sixth of its equivalent on Earth, the cable structure can be almost as light and as flimsy as gossamer. All these aspects suggest the most marvellous advantages. There will be no wet air to corrode the metallic parts of the telescope or any of its associated machinery. It may be no exaggeration to predict that from the Moon we shall see tens of times further into the depths of the Universe than has ever been possible before. Both radio and optical astronomy will be reborn on the Moon. From a world without clouds or atmospheric turbulence we shall see stars and galaxies hundreds of thousands times fainter than the faintest now visible from Earth. The furthest galaxies now visible from Mount Palomar, for instance, are about five billion light-years away, in the constellation of Coma Berenice, or Berenice's Hair.* On a photograph enlarged until the very grains of the emulsion are visible, we can see these galaxies as the minutest of specks. They are so faint and tiny that we need a magnifying glass to reassure us that they are not faults in the plate.[6] Yet the largest of these specks is believed to be a giant galaxy of a trillion stars, a galaxy ten times bigger than our own Milky Way! From future observatories on the Moon, we shall be able to examine the distant galaxies in Berenice's Hair in almost the same detail as we today examine the great spiral galaxy in Andromeda, which is 2,000 times nearer.

The mirror of Mount Palomar has a diameter of 200 inches, or nearly 17 feet. But on the Moon, where the natural factors of weight and stress that have limited Mount Palomar's mirror to this size do not exist, it will be possible to construct reflecting telescope mirrors with diameters of 2,000 inches, or 167 feet. And

* Berenice, wife of Ptolemy Euergetes, King of Egypt (247–222 B.C.) vowed to sacrifice her hair to the gods if her husband returned home the conqueror of Asia. She suspended her hair in a temple, but it was stolen on the first night. The priests appeased the furious queen by assuring her that the wind had carried her hair to heaven, where it still today forms seven stars near the tail of Leo.

the Lunar environment presents astronomers with a still greater advantage; the Moon rotates some thirty times more slowly than does the Earth. This means that time exposures of any desired point in the sky can be correspondingly longer. Mount Palomar's exposures are interrupted each morning by the coming of daylight. On the Moon there will be no such hindrance. Periods of up to 336 hours, or 14 earth-days, can be spent gathering a single exposure.* In the Moon's polar regions, where there is perpetual night, exposures can last *for a year or more*. With these opportunities and all the new inventions which will result from them, astronomy will surely undergo an improvement proportionately even greater than that introduced so long ago by Galileo.

The migration of astronomers to the Moon will not be the first occasion on which science has preceded industry. The entire impetus of the Apollo programme, after the first landing, was scientific. I once listened to a group of NASA engineers and scientists fiercely disputing the priorities for a forthcoming Moon mission. An engineer nervously suggested that one of the proposed scientific experiments should be dropped from the mission to make way for an extra back-up safety device. There was an aggressive rustling of papers from the scientists as they threatened to walk out and hold an indignant press conference. The proposal was quickly dropped, and the mission flew without that extra safety device.

The engineers knew that without scientific backing they would get no funds for exploration of the Moon. Yet it is often forgotten in these days of U.S.–Russian cooperation in space that the initial space programmes, from 1957 until about 1961, were inspired, not by science—most scientists at the time showed little interest in them—but by military fears. Funding of the American Vanguard and Mercury programmes, long desired in vain by

---

* All this emphasis on photography requires an explanation. A big telescope is virtually useless without cameras. Our eyes can only take feeble snapshots. We can sit for hours in a badly-lit room, and the brightness around us remains feeble. But an ordinary cheap camera, set with a 30-minute time exposure, can bring out every detail as if there were bright sunlight. An hour's exposure on a large telescope can distinguish stars 100 million times fainter than the faintest visible to the naked eye on a perfect night.

would-be space colonizers like Wernher von Braun, was mainly a reaction to Sputnik. The flight of Sputnik had aroused a deep fear that Russia would place nuclear weapons in orbit. Even the later Apollo programme was inspired by a desire for political prestige, which is perhaps only different in degree from the earlier fears of a military technology 'gap'. This progression from apparently urgent military necessity to pure scientific exploration, as the military threat recedes, is likely to be repeated on the Moon in a more subtle way.

It may be thought somewhat optimistic to assume that governments will happily pay enormous sums to install observatories on the Moon. One can almost hear people muttering angrily about the relative importance of stars and sewers, and disputing which ought to have financial priority. The scientists are wise enough now to know how long it would take them to win a battle against slogans like these. Wernher von Braun lost a similar battle in the 'fifties (his plans to launch satellites were rejected) *until* Sputnik, which seemed to pose a military threat. It is probable that pure scientists will always be at a disadvantage if the battle is fought under these conditions. When there is no political or economic reason to spend big sums of money except to learn more facts about the Universe—then governments will be reluctant to spend the money. But scientists have learned wisdom since the days of von Braun's frustrations, and they have discovered one rule of scientific politics, that in the absence of any menacing modern equivalent to Sputnik, it will be necessary to invent one.

The pure scientist has one great ally when it comes to lobbying for funds for outlandish projects. This is the Military Bureaucrat, a person whose ideals he can afford to despise in prosperous days, but whom he has learned to court when funds are thin. The archetypal military chief has no interest in pure science, and very little in sewers. But he has two great virtues from the point of view of the ambitious scientist. He is highly excitable, and he has great lobbying muscle. The following scenario suggests how an astronomer, determined to examine more closely the galaxies in Berenice's Hair, might set about obtaining Congressional funds for his Moon observatory. My scenario is very crude, and no doubt the actual process will be much more subtle.

Imagine a club in Washington, which is frequented by both civilian and military scientists. Our galaxy-enthusiast enters with two colleagues. Their fairly loud conversation at the bar goes something like this:

'I see we're making great progress in lasers.'

'The Russians are streets ahead of us. They're going all out to develop a high-powered carbon dioxide laser. And their army is funding the research.'

'That's bad news. The carbon dioxide laser could be the strategic weapon of the 'eighties. If it was effective up to several hundred miles, it could wipe out incoming missiles like a blowtorch. Any nation that possessed it would be free to commit aggression.'

'Wait a minute. They couldn't keep a thing like that secret. They couldn't test it on Earth without our spy satellites spotting it.'

'Who says they have to test it on Earth? They could test it in space and no one would know. Come to think of it, the Russians are planning to build a radio telescope on the Moon. That would be the perfect cover for testing the laser, since the operation of all modern radio telescopes includes laser technology.'

'Gentlemen, I must say I can't understand how those responsible for our national security can sit back and do nothing in the face of this threat.'

This conversation, or anything like it, should not be taken literally. But this kind of propaganda, if disseminated skilfully and persistently among the military hierarchy, preferably with its authors half believing in it, could easily stir a demand for funds to build a parallel Lunar radio telescope to counter the real or imagined 'laser threat'.*

The laser-weapon I have imagined is perfectly feasible, and, if perfected, would be much cheaper and more reliable than the anti-ballistic missile systems now being built. Even if the Russians had no plans to test weapons under cover of a radio telescope, or

* Dr Thomas O. Paine, former administrator of NASA, may have been toying with a propaganda exercise of this sort when he pointed out in 1969 that it would be possible to use a giant Solar reflecting mirror on the Moon, which could destroy any chosen city of Earth by fire. There is no evidence that any government at the time contemplated the use of such a mirror.

even if they had no plans to build a radio telescope at all, the American generals might think it prudent to go to the Moon and build one, well, just to remove any temptations. Military leaders feel a tremendous compulsion to keep up with the most modern military technology. Dr Strangelove may have been a maniacal comic figure, but he was only funny because with his obsessions about technological 'gaps' he was aping the eternal fears felt by Pentagon and Russian generals. The weapon that arouses these fears in the 'eighties will not necessarily be the high-powered laser, although many people think it likely since it is the logical successor to the surface-to-air or the air-to-air missile.[7] But it is certain that there will be such a weapon, and that it will be desirable to develop it on the Moon to hide its tests from prying eyes. Military technology never stands still, however much we imagine after each great breakthrough that we have the ultimate weapon. And I predict that the need for this technology will be seized upon by the astronomers as an excuse to take their telescopes to the Moon. They may, of course, return to the Moon without help from the military lobby; but this would indeed be an extraordinary achievement, since it would defy one of the dominant social trends of the age.*

This stumbling block overcome, and I have little doubt that it finally will be, let us look some twenty years further into the future. Men will not be content for long to use the Moon only for pure science and military experiments, for the very good reason that its environment offers great commercial and industrial opportunities. One has only to turn the pages of any issue of the *Journal of Vacuum Science and Technology* to understand the frustrations of terrestrial vacuum engineers, and to realize the great opportunities for manufacture which free and total vacuum would provide, since vacuum chambers on Earth must of necessity be small in area, imperfect, and expensive to construct.[8] Neil Ruzic, editor of the American journal *Industrial Research*, asked 1,742 members of the American Vacuum Society which devices or materials they believed could be produced more easily or better if production were done in a Moon factory.[9] The

---

* At least it will be difficult to prove me wrong. Secret projects usually remain secret. No Moon-based astronomer will dare to admit that the obliging fellow who changes his lenses is in fact a disguised military agent.

answers, with percentages of replies in favour of each product, were:

Vacuum-cast alloys, 70 per cent.
Vacuum welds, 56·8 per cent.
Electronic optical devices, 28·4 per cent.
Other optical components, including lenses, 17·5 per cent.
Medical and biological products, 13·3 per cent.
Industrial chemicals, 6·7 per cent.
Petrochemicals, 4·1 per cent.
Plastics, 3·3 per cent.
Miscellaneous, 5·8 per cent.

Many gadgets now used on Earth, from the ordinary light bulb and thermos flask to the more complicated television set or computer, depend on vacuum technology. A large number of industries making similar products to these require vacuum chambers costing more than three dollars per cubic foot, the best of which are only 80 per cent efficient. The *Journal of Vacuum Science and Technology* is full of reports about new, expensive apparatus for detecting leaks and maintaining high pressure in these chambers. All such elaborate machines will be unnecessary on the Moon. A worker will need merely to step outside on to the surface to find the most perfect of vacuums.

One can barely imagine some of the industrial possibilities of this situation. Adhesion of metals, for instance, which on Earth requires complex welding techniques, will be achieved on the Moon at no direct cost at all. It is either very difficult or impossible to weld together on Earth some of the newer, tougher metals. On the vacuum of the Moon, most of them need only to be touched together to make a perfect weld. What strange new metallurgies and engineering techniques will natural vacuum and low gravity introduce? Many of them have been predicted: improved optical lenses and mirrors, unique alloys and metal products with superior properties of weight, strength and purity, perfect ball-bearings and spheres, single crystals of larger size, higher purity and higher crystallographic perfection for electronic machinery, and base materials for advanced semi-conductors.[10]

There is tremendous promise also for biological and medical products. Perfect, or at least much more efficient, vaccines and antibiotics can be manufactured on the Moon or in orbital factories. Serums and viruses can be prepared in near weightlessness with none of the impurities which make our present immunization techniques so unreliable.[11] And the Moon will be an excellent place for *direct* medical care. People with weak hearts will live much longer in the weaker Lunar gravity. There will be relief for burn victims and people with muscle diseases, arthritis and rheumatism. We can predict a great demand for Lunar hospitals and old people's homes. When the late J. B. S. Haldane, the biologist, was dying of cancer, he wrote an impassioned letter to this effect to Arthur C. Clarke. What a boon, Haldane declared, would be a low-gravity hospital to patients like himself! 'I and a million other surgical cases would be quite satisfied with Lunar surface gravitation.'[12] His letter prompted Clarke to comment in one of his essays, 'So I get pretty mad when I hear ignorant but well-intentioned people ask, "Why not spend the space budget on something useful—like cancer research?" '[13]

The scarcity of metals is a problem on Earth. Landscapes are increasingly desecrated by mining operations; mining in the United States alone produces a billion tons of inorganic waste each year. Within the next 150 years, it is possible that we may exhaust all supplies of such essential metals as iron, copper, chromium and nickel, not to mention lead, zinc, silver, mercury, bismuth, tin and cobalt.[14] This is not to say that these metals do not exist in abundance in the Earth's crust. But beyond a certain point, which cannot be far off, it will not be possible to extract them without doing intolerable damage to the environment. They could always be mined by nuclear explosions, but anyone who even contemplated such a violent scheme on Earth would risk arraignment as a public enemy. The same difficulty applies to any large-scale mining of the oceans. The necessary deep-sea technology, requiring resistance to tremendous submarine pressures, would be even more costly and difficult than that demanded by space, and it appears unavoidable that effluent and waste would have to be pumped back directly into the sea, reducing the oxygen content and hastening the death of all marine life-forms. Yet these objections to mining on a huge scale do not apply on the Moon. Whole sections of the Lunar surface the size of Europe can be vigorously mined without doing any harm to anyone's environment. Wastes will be no problem; they can be left lying on the surface. Such projects are, of course, more than a century into the future, since their cost today would be more than any nation could afford; but it is certain that ultimately they will be carried out, since the alternatives will be the ruin of the Earth's environment and the collapse of civilization. Indeed, in future ages it will seem astonishing that the Earth was ever considered a unique place either for manufacture or for the extraction of raw materials.

It will be very much cheaper to transport goods from the Moon to the Earth than from the Earth to the Moon, since the return journey will always demand 97 per cent less energy. This is both because of the one-sixth Lunar gravity, and because the Moon's mass is one-eightieth that of Earth. All planets and all moons are at the bottom of what may be called their 'wells of gravity'. It is as if, to lift a payload from a planet's surface into space, we had to employ the amount of energy proportional to the planet's mass and radius.[15] To put it simply, to escape from the Earth we must travel at 7 miles per second, while to escape from the Moon 1·5 miles per second is sufficient. But because the Moon's surface gravity is so much weaker, we need much less energy to accelerate to 1·5 miles per second there than we would need to reach the same speed on Earth. Now the Earth's escape velocity of 7 miles per second is 4·7 times that of the Moon. The energy needed to travel into space from the Moon is therefore roughly one-fifth times one-sixth, or one-thirtieth that required to launch the same payload into space from Earth. This works out at an energy reduction of some 97 per cent. Nor is there any problem in transporting goods across the seemingly huge distance of 238,000 miles between the Earth and the Moon. Having once achieved Lunar orbit, a ship need only fire its engines for a few moments in order to break free from the Moon and travel in any direction it wishes.

I have heard some people say, 'Think how much it costs to propel a jet 3,000 miles from London to New York. Magnify that and you see how much more it would cost to transport goods back and forth to the Moon!' Even after the Apollo flights, some people have difficulty in grasping that in the vacuum of space, a spaceship will travel indefinitely at the speed it had reached when its engines were switched off—whereas a jet over the Atlantic, struggling against air resistance , will soon plummet into the sea if its engines fail. It is three centuries since Sir Isaac Newton published his First Law, that 'every object remains in the same state of uniform motion unless acted on by a force' (i.e. until braked by its retro-rockets). Yet whole generations of school-children seem to have graduated without ever having been taught this fundamental law of the Universe.

It may not be necessary to use rockets at all when launching payloads into space from the Moon. Arthur C. Clarke suggested in 1950 that an electric railtrack would in the long term prove the most efficient means of space travel from the Moon.[16] This hypothetical device has been named 'lunartron' by American space officials.[17] The lunartron track would be horizontal or gently rising, and would be two or three miles long, sufficient for the vehicle to attain the necessary high speed for launch. It would be unsuitable for *manned* vehicles (unless the track was at least 100 miles long) because acceleration would be too high. But containers of rocket fuel and other supplies could be shot from the Moon along the track and intercepted at any desired point in near-Earth or interplanetary space. It will therefore be cheaper to refuel and resupply Earth-orbiting spacecraft from fuel-dumps and warehouses on the Moon, since a vehicle has only to reach 5,000 m.p.h. to achieve Lunar orbit, compared with 25,000 m.p.h. to attain orbit from Earth. With these low-energy launchings, and its low gravity providing ease of spaceship construction, the Moon is also likely to prove convenient for launching missions to the planets. It will serve this purpose until the truly self-contained spaceship is invented, the fabulous ship that can travel from one planet's surface to another, without jettisoning a single part, and needing no facilities on its travels other than fuel dumps.[18]

A permanent or semi-permanent colony of Lunarians, number-ing some hundreds of thousands, is likely to exist on the Moon by the middle of the 21st century. They will find it safer and more convenient to live underground. If digging is carried out under the protection of a transparent dome (which will be essential for surface agriculture), underground living quarters, of any desired size, can be cheaply constructed in three basic stages.[19] A deep hole must first be dug by drill or explosives. A huge, deflated plastic balloon is inserted into the hole. It is then inflated from the surface by a pump, thereby creating a large, semi-spherical underground chamber with rigid walls. It will then be a simple final step to sub-divide the chamber into separate rooms. Many of these can be interconnected by tunnels, making, in effect, a sub-surface city of any desired size.

It is probable that by the year 2050 anyone on Earth with binoculars will be able to see the lights from the tops of these encampments on the darkened surface of the night-time Moon, shining like stars in places where no stars ought to be, the lights of man's second planetary civilization. 'Civilization' will be the correct name for it. Wernher von Braun believes that a baby will be born on the Moon during this century.[20] As he and many other space writers have remarked, such children, having been reared on one-sixth gravity, and equipped with the brilliant mechanical devices which their way of life will demand, will never want to live permanently on Earth. The gravity of the home planet will oppress them as if they were six times their actual weight. Its political and economic systems will appear to them both chaotic and dangerous, and the presence of many more bacteria in the air will require so many inoculations as to make the visit to Earth a purgatory. The Moon will attract so many thousands of immigrants that ultimately even the environment of this planet, with a surface area the size of Africa, will begin to deteriorate. As Clarke suggests, 'two hundred years from now, there will be committees of earnest citizens fighting tooth and nail to save the last unspoiled vestiges of the Lunar wilderness.'[21]

6

Venus, The Hell-World

Colonization of the Moon and the establishment of Lunar settlements on a huge scale will inflame rather than satisfy man's urge to expand his activities ever further into space. Although developments on the Moon will enrich our industries, restore our self-confidence as a race, and partly diminish our fears of premature extinction, the Moon will never be wholly satisfactory as a second world. Whatever changes we make on its surface, it will still be apparent to a Moon settler that he is living on a barren rock. The environment will be dull, and the discipline will be wearisome. Every sunset will look exactly like every other. The stars will glitter like a billion diamonds, but they will look exactly the same one month as they looked the month before. A man will be unable to leave his base unless he is wearing the most ridiculous clothes, in which the slightest tear could bring an unpleasant death. While he is outside his base or city, he must live by his watch. If he forgets to consult it, or if it goes wrong, he will run out of oxygen and die. There will be few sports or amusements, other than exploration and ball-games played in the weirdness of one-sixth gravity.

For safety and efficiency, life will be highly regulated. A man who even stops to marvel at the wonders of technology that sustain him on the Moon will imagine that he hears a small voice saying, 'Oxygen costs so many pence per cubic centimetre; you're here to work, not to enjoy yourself.' The huge blue Earth in the sky, which to the early astronauts seemed unbelievably beautiful, will to the colonist gradually come to seem hateful. Remaining always in roughly the same place in the sky, it will shine like a great baleful eye, envious and meddlesome. It will remind him always that he depends on the caprices of its markets and the penny-pinching of its accountants. He will feel that he is liable to be summoned home at any time because the demand for matchbox-sized computers has mysteriously fallen, or because his Earth-based company has a thin order-book and is cutting its investments. The Earth–Moon industrial system will probably last for centuries, but few settlers on the Moon in these early decades will feel this sense of permanence.

A minority of people on Earth will share these anxieties in a different way. The conquest of the Moon will to them be a triumphantly successful experiment—but an experiment only. To these people, it will have served to open the way to a much greater enterprise. Imperceptibly at first, but ever more loudly, a new idea will be expressed. There will be such talk as has rarely been heard since the Puritans of the seventeenth century became discontented with the oppressions of Stuart England, and decided to emigrate to America. Much simplified, it will run something like this:

'Every day, the Earth seems to be a slightly more horrible place than it was the day before. We scan the newspapers daily to see whether the balance of power has changed, and that nuclear war is imminent; or whether rebellion has broken out, and that honest citizens are being murdered; or whether extremists have taken over the government; or whether an elected statesman has been dynamited; or whether strikers have caused a shut-down of electricity or whether we are to die prematurely from poison because farmers have sprayed too much of the wrong kind of pesticide; or whether on the contrary, food prices have jumped by 50 per cent because the farmers are using insufficient pesticide; or whether the population has grown so large that people are going to starve on a scale never before imagined; or whether some zealous environmental scientist has discovered a new reason why all life is bound to end by the year such-and-such. Our predominating emotion is no longer excitement or pleasure, but fear sitting behind us like an ever-darkening shadow.

'It is time for us to leave this decaying world and find another. We must cut off our roots and build a new civilization that is utterly independent of the old one. Did not such migrations in the past lead to great flowerings of culture? Did not the followers of Moses, the Norman conquerors, the Elizabethan explorers, the Chinese merchants, the Huguenots, the Mormons, by leaving the lands where they were persecuted and despised, all attain heights of wealth and achievement that had no parallel in their former state? We must do the same before we too are caught up in Earth's ruin. It will not be difficult to persuade some government or a syndicate of industrialists to build spaceships for us in exchange for a share in the enterprise. On our new world there will be no wars and no pollution. We will be determined to prevent them. For the first few generations at least, there will be no excess of population. We shall build that new Eden that idealists have always yearned for, even though we may have to suffer years of misery and hardship. Where, then, shall we find it? The Moon does not interest us, since it is dominated economically by Earth. And we want to breathe natural air. We want meadows and forests and great fields of agriculture. Those stuffy Lunar domes and sealed caves are not for us. Mars has very little air of any kind. Let us go, therefore, to the planet Venus!'

Venus? An astronomer's first reaction would be incredulity that anyone could be so ignorant as to contemplate living there. Although Venus is almost as large as Earth, its present environment is totally unfavourable to human life. The proportion of water vapour in its atmosphere is about 0·7 per cent. There is a low proportion of nitrogen and, on an even smaller scale, compounds of mercury and chlorine. The rest, about 90 per cent, consists of the unbreathable gas carbon dioxide.[1] The Earth's atmosphere is only about 0·03 per cent carbon dioxide, but because Venus is considerably nearer to the Sun than is the Earth (67 million miles compared with our 93 million) carbon dioxide has accumulated remorselessly from volcanic sources, instead of being locked up in the planetary crust, as is the case on Earth. There was never any rain to prevent the growth of carbon dioxide at an early stage. From its accumulation has developed a process known as the 'Runaway Greenhouse Effect'. The temperature is higher inside a greenhouse than outside, because the Sun's radiation enters but cannot escape. Venus imitates this process on a planetary scale. The Sun's visible radiation penetrates the thick carbon dioxide atmosphere. As the planet's surface absorbs the Sun's long infra-red rays, it heats them up and re-radiates them. But they are then trapped by the clouds and atmosphere. Data from the unmanned American and Russian spacecraft that examined Venus in the 'sixties have given us the generally accepted conclusion that the temperature at the surface of Venus is approximately 900 degrees Fahrenheit (480 °C), which is twice as hot as the hottest part of the standard kitchen oven. Large areas of it, astronomers have reasoned, must literally be red-hot.[2]

The rich carbon dioxide atmosphere is responsible for tremendously high atmospheric pressures at the surface. A man standing on the planet, assuming that by some means he avoided being fried or suffocated, would be subjected to the same heavy pressure on his body as is a skin-diver at the uncomfortable depth of 250 feet under the sea. The atmospheric pressure is about 100 times greater than it is at the Earth's land surface. Now, assuming further that the man survives these various unpleasant hazards, and that his eyes can miraculously penetrate the thick dust and the low clouds, he would be in danger of being driven mad by extraordinary optical illusions. The high atmospheric density produces 'super-refractivity', in which light rays, instead of bending very slightly, as they do in water, actually bend by more than 90 degrees. A man standing on Venus would see the *entire surface of the planet*, including landmarks on the opposite side of it, because light rays from them, instead of travelling in nearly straight lines, would describe almost complete circles. He would have the impression that he was standing at the bottom of a deep bowl, surrounded by high cliffs which were in reality the horizons. It is difficult to imagine this in terms of Earth landscapes. In Venus conditions, a man looking south from Westminster Bridge with a telescope would see Canterbury Cathedral towering above him, some way up a gigantic cliff. Higher up and beyond it, he would see the Channel and above that the Mediterranean. But still higher and further, he would see to his astonishment Canterbury Cathedral again! He would be able to see Australia, not one single Australia but dozens of Australias all scattered around at different points on the cliffside, high and low. Yet he would not be witnessing total chaos in which an infinite number of Australias were strewn about at random. A geometrist working with light curves would be able to predict where each Australia should be, and why. It would be as if, for each observer on the planet, the whole surface was unfastened and turned inside out, in the same way that an umbrella can turn itself inside out in a strong wind. The man on Westminster Bridge is in roughly a similar position to an insect sitting on top of an umbrella which has done precisely this. He could travel southwards without having to do any climbing and reach Canterbury Cathedral. He would still seem to be at the bottom of a deep bowl, except that the geography of the surrounding cliffs would have changed. Looking back, he would see that Westminster Bridge was now up the cliff. Beyond and above it would be Scotland, and then Westminster Bridge again, and so on.[3]

These extraordinary conditions are not too difficult to understand. Five planetary bodies, Mercury, Venus, the Earth, the Moon and Mars, in all probability evolved from the same cloud of dust and gas that must have been hurled out from the fast-rotating Sun more than five billion years ago. Superficially, they resemble each other more closely than they resemble any other planet in the Solar System. Yet their atmospheres are utterly different. While the Earth's atmospheric pressure is about 100 times less than that of Venus, it is at least 100 times greater than than of Mars. Mercury's is very much less than Mars if indeed Mercury has any atmosphere at all, and the Moon definitely has none. These figures may explain the mystery:

Planet	Average distance from the Sun (millions of miles)	Mass (Earth = 1)
Mercury	36	0·05
Venus	67	0·81
Earth	93	1·00
Moon	93	0·01
Mars	142	0·11

This table tells a simple story. Venus and the Earth have roughly the same mass, and so this cannot explain why they have such different atmospheres. But their similar masses do suggest that if the Earth was perhaps 10 million miles closer to the Sun than it is, then it too would have developed a 'runaway greenhouse' carbon dioxide atmosphere, surrounding a planet on which no life could have evolved.[4] If Venus, on the other hand, was about 15 million miles further from the Sun than it is, its atmosphere might be identical to Earth's and it could conceivably have intelligent humanoid inhabitants.* Oxygen atmosphere sufficiently rich to support life could never have evolved on Mars, Mercury or the Moon, no matter where these planets were situated. Their masses are too small, which means that their gravitational fields are too weak to retain such atmospheres. Mars, the most massive of these three, has a thin carbon dioxide atmosphere which was accurately predicted on the basis of its planetary mass long before the unmanned Mariner spaceships confirmed its existence. It appears that if a planet is to retain any atmosphere at all, it must be slightly more massive than Mercury but not necessarily so massive as Mars. But an atmosphere rich enough to support fairly advanced Earth-type plant and animal life can only evolve naturally on a planet that is at least twice as massive as Mars.

The planet's distance from the Sun is also of great importance. It must be in the Solar System's 'ecosphere', in other words within that narrow region roughly bordered by the orbits of Mars and Venus. Beyond the inner limits of this region, it would be too hot for Earth-type organisms to live, and outside this region it would be too cold. The Earth is comfortably inside the ecosphere, although we can see from the table that it was a close-run thing and that our planet was formed dangerously near the too-hot limit.[5] There is little chance, however, that anything could happen to change our orbit now. Such a change would require a collision with another planet or a sudden gravitational perturbation from some unknown mass. The chances of such a disaster actually occurring must be considered highly remote.†

Venus, as we have seen, was formed on the very edge of the ecosphere, and has too much atmosphere resulting from too much heat. This probably was not always the case. It has been con-

---

* The life-span of such creatures might be a decade or so longer than ours, because the slightly weaker gravity would place less strain on their hearts. They might also average a few inches taller than us for the same reason. A man who weighed 150 pounds on Earth would weigh only 130 pounds on Venus.

† An ingenious science-fiction film, in which this actually happened, alarmed many people in the days before the 1963 Nuclear Test Ban Treaty. Entitled *The Day the Earth Caught Fire*, it was the story of two simultaneous American and Russian multi-megaton hydrogen bomb tests, the combined power of which was so great that it had the effect on the Earth that a retrorocket has on an orbiting spacecraft. It altered the Earth's orbit and sent it falling towards the Sun.

vincingly argued that Venus started its existence with an environment very much like the Earth's. For a few hundred million years, the two planets would have appeared to be identical worlds, with similar primordial atmospheres of hydrogen, ammonia, methane and water.[6] An observer at this remote period might have felt safe in predicting that advanced civilizations would eventually evolve on both of them. Then, perhaps a billion years after Venus was formed, a fatal change began. Warmed by a much closer Sun, carbon dioxide began to accumulate in the atmosphere, slowly at first, but soon very rapidly. The result, four billion years later, is the dreadful world I have described.

Venus appears at first sight to present a dreadful prospect to would-be colonists. The men who founded the New England settlements faced rigorous difficulties and dangers. They were hard men who would put up with almost anything to build their new world. But even they might have quailed at the prospect of being simultaneously fried, suffocated, crushed, blinded and driven mad. In the words of Professor Carl Sagan, director of the Laboratory for Planetary Studies at Cornell University, 'Venus is very much like Hell.'

Sagan, a biologist as well as an astronomer, is one of the world's leading specialists on Venus. It is almost impossible to find an article on Venus in which his work is not cited. He has played a prominent part in evaluating the data that were sent back to Earth by the unmanned Mariner spaceships to Venus in the 'sixties. He has long been fascinated by the questions of life on other planets and by the origins of life on Earth. He was one of a group of biologists who confirmed the theory that it was ultraviolet light from the Sun that originally produced organic compounds from the Earth's primordial atmosphere. He made a laboratory simulation of the Earth's primitive, poisonous atmosphere and bombarded it with ultra-violet light. This action produced a set of compounds known as nucleotides and nucleosides, some of the basic building blocks of life. His semi-popular book *Intelligent Life in the Universe*, which he wrote in conjunction with the Soviet astronomer Iosif Schlovskii, is considered the current authoritative work on this vast but speculative subject.*[7]

* Because of Soviet restrictions, Sagan and Schlovskii never set eyes on each other until several years after the book was published. The composition

Sagan published in 1961 a truly extraordinary paper on Venus that may in time change human history.[8] The bulk of it was a fairly routine summary of man's knowledge of Venus to date, but in the last two pages he changed his approach and proposed one of the boldest schemes which man has ever contemplated. He had long brooded about the horrible environment of Venus, and had concluded that all the troubles of the planet, the heat, the pressure, the super-refractivity, all had a single cause—the rich atmosphere of carbon dioxide (CO_2). He suggested that all these troubles could be removed, and that Venus could be made as pleasant to live on as the Earth, if only the carbon dioxide could be split up into its two components, carbon (C) and oxygen (O_2).

On Earth, the breaking down of carbon dioxide into its two component elements is one of the normal links in the chain of photosynthesis. Plants and trees do it every second of the day. The carbon dioxide that we exhale is turned back into oxygen and carbon by our abundant plant life. If there was no plant life, and no oceans to produce the rain that nourishes that plant life, we would soon cease to have any oxygen to breathe. What Sagan proposes is simply this; if suitable plant life can be introduced into the atmosphere of Venus, exactly the same thing will happen there. The carbon dioxide will break up, making oxygen for the colonists to breathe. Sagan, and now many of his colleagues, believe they have found the ideal form of plant life to introduce into Venus's sky.

---

of each subject had to be agreed by mail. 'The probability of our ever meeting is unlikely to be higher than the probability of a visit to Earth by an alien,' Schlovskii wrote sadly in the Introduction.

7

Making it Rain in Hell

Some of the smaller Mediterranean islands lack natural springs of water. Many of them are too remote and too sparsely inhabited for water to be carried to them economically, and so the inhabitants are forced to collect and store their drinking water in rain-tanks. If this water is to be kept pure, a live eel must be placed in the rain-tank. For rainwater, like the air from which it falls, contains tiny micro-organisms, which the eel devours as he drinks the water. When the eel dies, an unpleasant smell from the rain-tank warns its owner that the micro-organisms are reproducing unchecked and making the water stagnant. The tank must then be washed out and refilled, and another live eel placed inside.

Micro-organisms such as these are the oldest and the most numerous of all living species on Earth. Although existing in hundreds of different varieties, they are classified under the general name of algae. They range in type from pond scum to the growths that form some kinds of giant seaweed. Of all known life-forms the algae micro-organisms seem the best fitted to live in hostile conditions where there are even the minutest quantities of water.[1] Some have been found breeding happily in the fuel tanks of jet aircraft, living in kerosene and oblivious to violent changes of temperature and pressure. Another group has been found in the cooling water that circulates through the cores of nuclear reactors, where a human being would be quickly killed by radiation. Some species grow in the barren wilderness of the Antarctic, where there is scarcely any soil because the ice is housands of feet thick, and the temperature is often below minus 100 degrees Fahrenheit.[2] Others have survived for years frozen in blocks of ice. Yet they appear also in warm springs, which are nearly 400 degrees hotter.

But of all forms of algae the blue-green, or *Cyanophyta*, is the hardiest. Blue-green algae were, with little doubt, the first genuine life-form—as distinct from a mere organic component—that ever lived on Earth. They are neither wholly a plant nor an animal, but are something of both. Although microscopic in size, they are distinguishable from bacteria (which are generally classified as plants), but similar in that their reproduction is asexual and very rapid, and that they lack a distinct cell nucleus. Yet they are invariably one-celled, like their descendants the protozoa, the microscopic organisms which biologists consider to have been the first of Earth's animals. Blue-green algae were the parents of bacteria and the parents of protozoa, and through them the parents of all the great variety of life-forms on Earth today.

Three billion years ago, the Earth's atmosphere had little of its present agreeable mixture of oxygen and nitrogen. Instead, it contained huge quantities of carbon dioxide, ammonia and methane.* Then, from shallow sunlit seas, the blue-green algae attacked the all-pervading carbon dioxide to obtain the carbon that would give them glucose and other carbohydrate foods.† The oxygen which had been liberated from the carbon dioxide swept through the seas and sky, destroying the ammonia and methane. This oxygen enabled animals to evolve which in turn exhaled more carbon dioxide and thus provided more food for plants, and so on, in an ever-rising spiral.[3] The blue-green algae were the direct ancestors of us all. Its hardiness, its tremendously rapid rate of reproduction, its easy availability and its lust for attacking carbon dioxide have persuaded Sagan and his colleagues that it is the perfect substance for injecting into the Venus atmosphere with the purpose of turning it into oxygen.

The genius of this scheme lies in its simplicity, its relative cheapness, and the short time, perhaps as little as two or three years, which it would take to produce results. A few dozen spacecraft, most of them unmanned, will be put in criss-cross orbits

* Carl Sagan considers it improper to call these gases 'poisonous' simply because they would poison *homo sapiens*. There may be creatures on other planets of such different chemistry that they breath ammonia and would be poisoned by oxygen. Sagan thinks it important to avoid 'oxygen-chauvinism' when considering extra-terrestrial life.

† Despite its present-day familiarity in chemists' shops, glucose appears in nature as the most basic of carbohydrate foods. Its chemical composition is six atoms of carbon, twelve atoms of hydrogen and six atoms of oxygen. It combines with sunlight to form oxygen and starch.

around Venus. Each spacecraft will carry on board large numbers of small, torpedo-like rockets. Every 90 seconds, at points separated by about 500 miles, each spacecraft will fire one of these rockets into the atmosphere. The nose-cone of each rocket will contain a colony of blue-green algae. A few ounces of T.N.T. will explode the nose-cone as soon as it has penetrated the carbon dioxide clouds, and the algae will start to feed and to reproduce. Once begun, this rate of reproduction will increase so rapidly, and the carbon dioxide will be so swiftly broken down, that perhaps within a year the surface of Venus will be partly visible to tele-scopes on Earth.[4] The algae's rate of reproduction is the key to the success of the plan. On Earth, the algae were at first so sparse that some two billion years elapsed between their first attack on the primordial atmosphere and the evolution of advanced animal and plant life. However rapid a geometric rate of increase (i.e. 1, 2, 4, 8, 16, 32, 64, etc.) may be, if the starting number of organisms is too small in relation to the size of their environment, it will take millions of years for the growth-rate to produce a dense population. But once the growth-rate has become critical, the numbers soon become colossal.* And so in the atmosphere of Venus, where the numbers of individual algae at the beginning will not be ones and twos, as was the case on Earth, but hundreds of billions, and where they can be at any time replenished or replaced by a hardier strain if their progress is unsatisfactory, the time-scale needed to provide a planet with life may be reduced by a factor of nearly two billion.

We can be almost sure that the blue-green algae will not only survive but proliferate in an environment of nearly pure carbon dioxide. Until 1970, nobody was absolutely certain that they would, since no experiment to test the theory had been made.

---

* A famous anecdote illustrates this idea. An eastern king was grate-ful to one of his subjects for a service performed, and asked him to name his reward. The man asked for the number of grains of corn that would cover the 64th square of a chess-board according to a certain formula; he asked the king to place one grain of corn on the first square, two on the second, and so on, doubling each time until the board was covered. The number of grains on the 64th square would be the reward. The king thought the man was easily pleased, and sent for a sack of corn. but he soon found to his horror that he needed far more corn than was available in his whole kingdom, the actual number, 2 multiplied by itself 63 times, being 9,223,372,036,854,775,808.

But four biologists in that year, with Sagan's plan in mind, carried out a series of tests to see how various algae cultures would fare in tanks filled with carbon dioxide.[5] To make their simulation of the Venus sky more realistic, they increased the atmospheric pressure in the tanks to the utmost that the walls could withstand. The results were not only a triumphant success, but improved Sagan's plan. Not only did the algae start to produce oxygen at a vigorous rate, but the rate itself showed a continuous increase. In one experiment typical of the series, it was shown that each million cells of algae were increasing the oxygen by 380 per cent each day. It was found that the most prolific and therefore the most suitable strain of blue-green algae for Venus was a single-celled genus found on Earth in hot springs and known as *Cyanidium caldarium*. If the quantity of oxygen in the Venus atmosphere can be increased by 380 per cent per day, it cannot be long before spectacular changes are apparent.

Soon will come what some science-fiction writers, excited by Sagan's plan, have called the 'Big Rain'.[6] As oxygen replaces the carbon dioxide, the Sun's infra-red radiation, hitherto trapped, will escape into space, and the temperature of the lower atmosphere will decrease considerably. Water will collect from the atmospheric vapour. The raw material for 100 inches of rain, spread evenly over the entire planet, exists in the Venus atmosphere. Eventually, torrents of water will lash down on the glowing, storm-swept surface where it has never rained.

The first cloudburst of the Big Rain will certainly not hit the ground. Rain just cannot fall on to a surface of 900 degrees Fahrenheit. While still at altitudes of thousands of feet, it will vaporize into steam and rise up again into the high atmosphere from which it came. But something will have been achieved by this first attempted encounter between cool water and extreme heat. The ground temperature will drop, perhaps by 100° F. Meanwhile, higher in the sky, the accelerating breakdown of carbon dioxide into oxygen will continue. Photosynthesis will occur as the algae make carbohydrates from the atmospheric oxygen and carbon. From these carbohydrates will come the complex organic chemicals that will bring plant life to the planet's surface.

Soon the Big Rain will try again. This time it will come nearer to the surface before it vaporizes, and the ground temperature will again fall, perhaps by another 150° F. After each attempt, the surface will be made cooler, and the carbon dioxide above will be broken up more thoroughly. At last, when the ground is about 200° F., the torrent will strike. The deserts of aeons will dissolve into muddy rivers and lakes under this thundering wall of water. Most of it will trickle away into the 'soil' (if a rocky desert surface can be given such a name), forming billions of tiny underground channels, and moistening the sub-terrain in preparation for the arrival of the complex molecules that will form the building blocks of plant life. The final conquest of the surface by the Big Rain will have knocked the temperature down still further, perhaps to 70 or 80° F. 'When more rain falls,' Sagan explains, 'the heat-retaining clouds will partly clear away, leaving an oxygen-rich atmosphere, and a temperature cool enough to sustain hardy plants and animals from Earth.'[7] Oceans will form in the depressions, perpetuating the cycles of rain that will nourish the plant life.

One last reaction will occur as the sky clears and the Sun can be seen for the first time in the long and monotonous history of Venus. The new atmospheric oxygen will combine with sunlight to create, high in the stratosphere, a layer of ozone, which absorbs dangerous ultra-violet Solar rays. Ozone layers are indispensable to a planet where people live without space-suits. They protect all life on Earth; without them, our world would be uninhabitable.* Sagan's entire scheme, requiring no greater expenditure than that involved in a dozen or so orbiting spacecraft, and a few thousand small algae rockets, would bring the staggering riches of a second world into our possession. It is the simplicity of the idea that makes it so attractive. Plans for the redesigning or 'terraforming' of neighbouring planets have been contemplated since before 1950. But these plans were extremely complicated and expensive. They either depended on plant life already existing on the planet in question, or else they required physical engineering

---

* It used to be fancifully predicted that supersonic aircraft, flying at 60,000 feet above the Earth, would disrupt the ozone layers and endanger all life below the aircraft's flight path by admitting ultra-violet light. This was actually one of the grounds on which the U.S. Senate voted to scrap the S.S.T. in 1971. But despite innumerable flights by space rockets and high-altitude aircraft, no such disruption has occurred. It is certain that ozone layers repair themselves immediately after being penetrated by a solid object.

on a planetary scale that would cost quadrillions of dollars.[8] The gross national products of the industrialized nations will simply be too small during the 21st century to permit any projects costing quadrillions or even trillions of dollars, however socially desirable they might be. By the 23rd or 24th centuries, it will be a different matter, as Chapters 12 and 13 explain. In those far-off times, the human race will find it necessary to embark on engineering projects requiring such financial outlay that their mere suggestion today would reduce a Bureau of the Budget to hysteria. But until the second half of the 21st century at least, a national or international space agency, or a consortium of corporations, will have to restrict itself to space projects that cost less than a trillion dollars—even allowing for the inflation that will by then have occurred. Carl Sagan's plan for Venus, with its modest fleet of spacecraft of which only a minority need be manned, can be financed well below these limits.

An ethical problem could arise if animal life is detected in the central atmosphere of Venus, about forty miles (some 200,000 feet) above the surface. This part of the atmosphere is very much cooler than 900° F. It is so much colder in fact that many astronomers believe that the central clouds contain large quantities of ice crystals.[9] The bottoms of these clouds are slightly warmer, and probably consist of water droplets. In the lower half of these clouds, Sagan has suggested, living creatures may exist.[10]

I do not mean to imply the existence of some vast, aerial civilization, for ever floating lazily in the clouds. Such creatures could be no more advanced than jellyfish. Sagan speculates that they might resemble 'gas-bags', and their size would be somewhere between that of a ping-pong ball and a football. They would propel themselves around in the sky by the same principle as a jet engine, by sucking in gases in front and thrusting them out behind.[11] They would be a form of animal life that lived on carbon dioxide. The algae, by breaking down the carbon dioxide, would be destroying their habitat, and in a sense man would be committing genocide. This moral dilemma was imagined in a famous work of science fiction in the 'thirties (science fiction is often invaluable for discussion of hypothetical problems), Olaf Stapledon's *Last and First Men*, in which he vividly extrapolated the future course of human history over billions of years. Stapledon's Venus was partly covered by oceans, which made its terra-forming very much easier. In this scene, an evolved human species known as the Fifth Men was compelled to make Venus habitable and to migrate there because the Moon was about to crash into the Earth:

> Another trouble now occurred. Several electrolysis stations on Venus were wrecked, apparently by submarine eruption. Also, a number of ether-ships, engaged in surveying the ocean, mysteriously exploded. Evidently there must be intelligent life somewhere in the oceans of Venus. Evidently the marine Venerians resented the steady depletion of their aqueous world, and were determined to stop it. As all efforts to parley with the Venerians failed completely, it was impossible to effect a compromise. The Fifth Men were thus faced with a grave moral problem. What right had man to interfere in a world already possessed by living beings? On the other hand, either the migration to Venus must go forward, or humanity must be destroyed.
>
> It was resolved to put the Venerians out of their misery as quickly as possible. The consequent vast slaughter of them influenced the human mind in two opposite directions, now flinging it into despair, now rousing it to elation. The horror of the slaughter produced a haunting guiltiness in all men's minds, an unreasoning disgust with humanity for having been driven to murder in order to save itself. But soon a very different mood sprang up. The murder of the Venerians was terrible, but right. It had been committed without hate. This mood, of inexorable yet not ruthless will, intensified the spiritual sensibility of the human species, refined, so to speak, its spiritual hearing, and revealed to it tones and themes in the universal music which were hitherto obscure.[12]

On the real Venus, as we have seen, there is very little likelihood of finding intelligent life. While the deliberate extinction of intelligent alien life in order to replace it with a human colony would rightly be considered an outrageous crime, the removal of billions of creatures possessing no greater mental powers than amoeba or jellyfish will certainly cause no crisis of conscience. Such outcry as there may be will come from scientists lamenting the destruction of a species before it had been properly studied,* a protest against the *timing* of the act of destruction rather than against the act itself.[13] The Venus colonists, however, will ignore even this milder tone of objection. They will be more interested in creating their own habitat than in preserving one that is hostile to all life.

Living on Venus will have its inconveniences. For reasons which are still not understood, the planet rotates extremely slowly, so slowly in fact that 118 Earth-days elapse between each sunrise.† Each day and each night lasts about 60 Earth-days. The two-month nights, with their extreme cold and icy blizzards, will resemble Arctic winters. To avoid these vigils, the colonists will need day-camps and night-camps, each at the antipodes. By keeping on the move in this manner, perhaps by rail or aircraft, from one camp to another, a journey of about 10,000 miles every two months, it will be possible to live in perpetual daylight. The slow rotation cycle might at first sight seem to discourage the building of large and elegant cities, whose fixed and permanent character, according to Lord Clark's theory of civilization, is always evidence of an enduring culture because it demonstrates the self-confidence of a race that builds for the future, in contrast to a tribe of tent-dwelling nomads.[14] But for this reason, because the settlers of Venus will want to demonstrate, if only to themselves, their confidence and their sense of permanence, I predict that such cities will be built, however inconvenient the seasons. Tolerable nocturnal cheerfulness can doubtless be provided by bright lights and heated streets. Synchronous satellites will no doubt eventually be provided which, by means of powerful nuclear reactors, will bring artificial daylight to any desired place.

I am trying to draw a picture of the early growth of a young world, peopled by adventurous farmer-colonists and others determined to be independent of Earth and to escape to Venus from the Earthly dangers of pollution, over-population and War. But even the new Venus civilization, with its elysian landscapes and huge agricultural pastures, cannot be expected to prosper for

---

* To the question, *when* has a species been adequately studied, any biologist worth his salt will answer, 'Never.'

† For unknown reasons, Venus also has a retrograde rotation. This means that the Sun would rise in the west and set in the east.

much more than two and a half centuries. Venus cannot be more than a temporary cure for claustrophobia on the home planet. The problems of Earth will soon re-create themselves. There can be no pre-industrial era of machineless bliss, since the colonists will arrive armed with the full powers of 21st-century technology. They will not be in the least interested in birth control, since they will at first occupy and exploit only a tiny part of the vast surface area of the planet,* and also because a fast-growing population will seem to them the best insurance against any decision by governments on Earth to recall them home. Very probably, their attitude towards family planning will be the opposite of ours today. With the repeated slogan of their leaders, 'Let us populate our world,' most people will consider it their duty to marry in their late 'teens and have eight or nine children. A man can expect to be a grandparent before he is 35, and a great-grandparent before he is 55. Assume, for instance, that the colonists originally number a thousand, are of fairly equally mixed sexes, and possess such advanced medical skills that fatal disease is extremely rare. Assume that they decide on a population policy of maximum growth, for the reasons I have given above. Their average annual growth-rate of population could easily be as high as 7 per cent after a few years. This would mean a doubling every 10 years. It can be calculated that, at this rate, in two centuries the original population of a thousand will have reached a billion, and that in three centuries it will be 1·1 trillion, more than 330 times the Earth's population in 1970!

However noble the intentions of the colonists at the beginning, the natural human tendency to quarrel will soon assert itself. They will quarrel with Earth when they are few, and they will wage civil wars against each other when their numbers become great. Within a few hundred years, the dream of occupying a new world of huge open spaces will have died, and man will be the proprietor of two worlds, of almost equal size, and each plagued by similar problems.

Unfortunately, Venus is the only other planet in the Solar System whose size and proximity to the Sun makes it suitable for comparatively cheap terra-forming. The atmospheric carbon di-

* The surface area of Venus is 182 million square miles, compared with Earth's 197 million.

oxide of Mars, as I have explained, is about one ten-thousandth of the density of Venus's, and so there is little hope of profitably seeding it with algae. Sagan has proposed a scheme for terraforming Mars, which involves melting its north polar ice-cap and distributing the resulting water and vapour across the surface of the planet.[15] But this plan requires considerable engineering activity at the surface, and would therefore be much more difficult and expensive than his Venus plan.

As for the big planetary masses in the Solar System, it will take many centuries and great expense before they can be exploited for the actual *building* of new, Earth-sized worlds close to the Sun, a development discussed in Chapters 12 and 13. It will be necessary in the meantime to migrate to the planets in orbit *around other stars*. Yet the stars, as distinct from the Sun's planets, are so far away that the chances of men ever reaching them might at first consideration seem very slight. The nearest star, Proxima Centauri, is no less than 24 million million miles from Earth. This huge distance can perhaps be more easily visualized if we imagine the Sun contracted to a mere foot in diameter. If this Sun was situated in the middle of Piccadilly Circus, Earth would be a tiny ball 107 feet away. The giant planet Jupiter would be 560 feet away, a mere 1·2 inches in diameter, and Pluto, the most distant of the known Solar planets, would be just under a mile away, at Hyde Park Corner. But Proxima Centauri, by this reckoning, would be more than 4,000 miles away, somewhere near Kansas City. At the present maximum speeds of Apollo spacecraft, 25,400 m.p.h., it would take many centuries for a spaceship to reach the nearest star. Even if such craft could be accelerated to hundreds of millions of miles per hour, the voyage-times would still have to be reckoned in years. Fortunately, the 1960s have seen a great increase of cosmological knowledge. Our assessments of the physical nature of the Universe have radically changed. The next two chapters will show how this new knowledge gives us hope that spaceships may one day visit other stars by taking short-cuts through the immensities of interstellar space.

8

The Astronaut's Shrunken Head

Sir Isaac Newton brooded for many years on the problem of why an object falls to the ground. It was obvious that a message was transmitted in some way from the ground to the object. But through what medium? Newton was never able to solve this mystery, and even Albert Einstein three centuries later never settled it to his complete satisfaction. Clearly it was not air which transported the message, since Newton knew from his calculations on the dynamics of moving bodies that objects on an airless planet such as the Moon would fall just as surely. This riddle was an irritating flaw in his all-embracing laws of universal gravitation. Everything else seemed to fit into place. The entire Universe was held together by mysterious and invisible gravitational forces. The planets revolved around the Sun in orbits of the most meticulous harmony. Newton's successors discovered that the Solar System itself revolved around the centre of the Galaxy, returning to its starting point every 200 million years, or one 'solar year'.

But the fact which still bewilders us today is that space is virtually empty.* It contains, on average, about one hydrogen atom per cubic inch; the rest appears to consist of absolutely nothing. How, then, can it be possible for gravitational waves to transmit themselves from one point to another? They cannot travel through a medium consisting of nothing. To Newton and Einstein the question was very interesting, but of academic interest only. But in the next century it will decide the future of the human race. We shall learn from experiments dealing with the nature of space itself whether it is possible to travel within

* I refer to space *in general*. Dense hydrogen clouds can be found in certain local regions. Astronomers have also found traces of formaldehyde, carbon monoxide and other organic compounds. But these have nothing to do with the actual transmission of gravity.

reasonable voyage-times beyond the Solar System to the stars.

An engineer today, if given enough money, could design a star-ship without having to bother himself with these obscure questions. But his ship, because of the vast distances of interstellar space, could only reach even the nearest star after a voyage of intolerable duration. Its speed would be limited by the absolute maximum of 670 million m.p.h., the speed of light in a vacuum, which Einstein showed in his Special Theory of Relativity of 1905 to be unattainable by any material object. As it approached that speed, its length in the direction of motion would shrink towards zero, and the energy required to propel it would rise towards infinity. Although, as we shall see, a radical new interpretation of Relativity has come into existence today, the ortho-dox relativists say simply this: that since no spaceship can exist that is of zero length and that possesses an engine of infinite power, the limiting speed of light must mark an impassable barrier to man's expansionist ambitions. In short, they regard interstellar flight as an absurdity. As Edward Purcell of Harvard vigorously remarked, 'all this stuff about travelling round the Universe in space suits, except for local exploration, belongs back where it came from, on the cereal box.'[1]

But, happily for mankind, space is today proving to be a very much more complicated area than even the Special Theory describes. The late J. B. S. Haldane prophesied that we should never be able to understand the Universe. 'I suspect', he wrote, 'that it is not only queerer than we imagine; it is queerer than we *can* imagine.' The Special Theory alone appears at first sight to describe a Universe of utter chaos and madness. Two couplets express this vividly:

> Nature and Nature's laws lay hid in night:
> God said 'Let Newton be!' and all was light.
> It did not last. The Devil howling, 'Ho!
> Let Einstein be!' restored the status quo.[2]

The strangest conception which Einstein introduced in 1905 was that of 'time dilation'. His ideas seemed to defy all common sense; he himself dismissed common sense as 'a deposit of prejudice laid down in the mind before the age of eighteen'.[3] With one bold series of four equations he overthrew Newton's conclusion that time everywhere moves forward at a constant pace of one hour per hour. Astronauts in a spaceship travelling away from Earth at 99 per cent of the speed of light would age seven times more slowly than people on Earth. The length of their spaceship would contract by a factor of seven, and the engines required to maintain their acceleration would have to employ seven times more power. This was not an illusion to be explained by psychologists; the astronauts' measuring rods and the very watches on their wrists would record these altered times and lengths. They themselves would notice nothing peculiar. To them, if they possessed telescopes of phenomenal power, everybody back on Earth would seem to be ageing much more slowly. This would be explained by the fact that since the astronauts were living seven times more slowly, they would process their information seven times more slowly. They would discover their error on returning home. They would find that during their voyage the people on Earth had aged not seven times more slowly but seven times more rapidly! If, during their voyage, they succeeded in coaxing their stubborn engines to still greater speeds, to more than 99·9999 per cent of the speed of light, they would return to find that Earth had aged, not seven times, but *millions* of times more rapidly than they.

At the speed of light itself (assuming this was possible to achieve), the astronauts' time would stop altogether. Their journey would be instantaneous. And beyond the speed of light, their time would run backwards. This impossible feat has inspired the following limerick:

> There was a young lady named Bright,
> Who travelled much faster than light,
> She started one day
> In the relative way
> And returned on the previous night.

She must therefore have encountered a duplicate of herself before setting out. But this would be absurd. There can never have been two Mrs Brights since the original lady set out with no memory of having met her duplicate the night before. A sequel points out that a continuation of these faster-than-light voyages would produce an infinite number of Mrs Brights:

> The lady was Bright but not bright,
> And she joined in next day in the flight;
> So then two made the date,
> And then four and then eight,
> And her spouse got the hell of a fright.[4]

In all these ridiculous adventures, the law of 'causality' is being violated. Causality forbids any faster-than-light journeys through the known Universe since a consequence cannot occur before the event that caused it. Otherwise, we would have to assume the existence of an infinite number of parallel worlds running on different tracks. This is a fascinating idea, but it runs into all kinds of logical impossibilities.

Returning to the theoretically possible, the Special Theory is best illustrated by a thought experiment known as the Twin Paradox. One of twin brothers stays behind on Earth, while the other becomes an astronaut, setting out at the age of thirty for Planet X in another star system at 99 per cent of the speed of light. The traveller returns after five years to find that his twin is sixty-five—five of the space-twin's years will equal thirty-five of the Earth-twin's. It is meaningless to say that one time is right and that the other is wrong. Both are right. The rate of the passage of time depends on the speed through the Universe of the clock that measures it.

The whole fantastic mathematical structure of the Special Theory rests upon the still unexplained behaviour of light. While Einstein was a child, two American physicists, Albert Michelson and Edward Morley, found by a series of experiments that light rays in the vacuum of space move at a constant speed of 670 million m.p.h. *irrespective of the speed of their source.* Suppose that a star was rushing towards us at half the speed of light. We would expect its rays to reach us at one and a half times that speed. But Michelson and Morley proved that this was false. Whatever the speed of the light source, whether it was approaching or receding, the light always arrives at precisely 670 million m.p.h. If two cars, both travelling at 50 m.p.h., were to collide head-on, the speed of their collision would be 100 mp.h. But no such statement would be true of light rays. These do not collide head-on at twice the speed of light, as common sense would lead us to expect. They meet at precisely 670 million m.p.h. How, then, does time contract at high speeds? It is obvious, in view of the behaviour of light. An astronaut racing through the cosmos at half the speed of light measures the impact speed of a light ray from a star directly ahead of him. It reaches him at 670 million m.p.h., when because of his own speed towards the star it ought to be reaching him much more rapidly. He reasons that his measuring instruments must be at fault. In a sense he is right. He can only draw one conclusion; *his clocks are running more slowly*. He might then experiment further. Suppose that he climbs out on to the hull of his spaceship and measures the time it takes for the star's colliding light beam to travel from the bows of his ship to the stern. He might think that the beam would move at one and a half times the speed of light, because of the ship's speed towards the light-source. But it does not. In fact, the beam traverses the length of the ship at exactly the speed of light. The only possible conclusion which the astronaut can draw is that the length of his ship has shrunk in the direction of motion. There is, nevertheless, no way he can confirm this by observation, since *everything* on the ship has shrunk in the direction of motion. His measuring rods are all altered. While he looks towards the bows of his ship, even the distance between the back of his head and the tip of his nose will have shrunk.

While making these measurements he passes a planet. The astronomers on this planet, with amazingly sophisticated telescopes, observe the astronaut and his ship as they streak past. But they do not see a normal-looking astronaut or a normal-looking ship. Instead, they see an incredibly thin man standing on the hull of a strangely squashed-up ship. The astronaut himself cannot sense anything unusual. If, while facing the bows, he measures the distance between the back of his head and the tip of his nose he will find it normal. How can it seem anything else when the actual ruler which he uses to make the measurement is itself shrunken in length? Visual observation will tell him nothing. Only his reasoning, based on his measurement of the star's light-beam, will prove to him that his ship has shrunk in length and that his clocks are running slowly. No matter how fast he runs into that light beam he can never collide with it at any speed greater than 670 million m.p.h.

An understanding of the Special Theory of Relativity is essential if we are to understand the more complicated General Theory that followed it in 1916, and which has led in turn to the strange science of 'geometrodynamics', which is beginning to reveal to us the existence of hidden paths in space. Through these paths, as we shall see, it may prove possible for a material object to disappear in one part of the Universe and reappear instantaneously in another, like the Cheshire Cat in *Alice in Wonderland*. By use of such a secret path, the object would in effect have moved much faster than a light ray, but by taking a short cut it would have avoided the restrictive speed limit of the Special Theory. Modern interpretations of the General Theory lead remorselessly to the conclusion that such astonishing journeys occur constantly in the Universe.

Einstein, unlike many of his admiring colleagues, was not satisfied that his Special Theory explained all the puzzles raised by the Michelson-Morley experiment. He brooded about the need for a 'unified field theory' that would describe the Universe in its entirety and explain every cosmic event. He pondered the dream of the Marquis de Laplace, the 18th-century French mathematician * who speculated on the existence of an intelligence

which knew at a given instant all of the forces by which nature is animated, and the relative positions of all the objects, and which, if it were sufficiently powerful to analyse all this information, would include in one formula the movements of the most massive objects in the Universe and those of the lightest atom; nothing would be uncertain to it; the future, as the past, would be present to its eyes.

Einstein's yearning to imitate that vast intelligence made him ask *why* the astronaut aged more slowly, *why* did light behave in this strange way, *why* did mass increase with velocity, why, why, why? It was only possible to conclude that all these things were directly caused by gravity, the power that has baffled astronomers through

* Laplace later became Minister of the Interior under the first Napoleon—by whom he was furiously sacked after six weeks for 'introducing infinitesimals into administration'.

the ages, and whose workings even today are but dimly understood.*

Some writers have ridiculed the Twin Paradox, declaring that the space-twin could not possibly age more slowly than his brother. Surely, they have said, the Earth may be said to be travelling away from the spaceship at the same speed at which the spaceship is moving away from the Earth. Why, they ask, should the Earth be regarded as a fixed frame of reference while the Spaceship is considered to be a moving object? Einstein's devastating answer is that when the Earth moves, *the entire Universe moves with it*, and that when the spaceship moves it moves alone. The Earth-twin does not move relative to the Universe, and the space-twin does. He has used his engines to *break away* from the natural orbit which gravity would otherwise have allotted to him. The mass of his spaceship is infinitesimal compared with the mass of the Universe, and it is therefore the gravitational field of the Universe which prevails, and not that of the spaceship. For the Paradox not to work, the spaceship would need a mass equal to that of the Universe; it would need to contain as much matter as that of many billions of galaxies. But since the spaceship is in fact likely to be very much smaller than the Earth, it must yield to the laws of a very much greater mass.

A small American town was particularly jealous of its independence from the state government, and it liked to assert this independence in a number of petty but harmless ways. When the state introduced a speed limit on all roads of 70 m.p.h., the town agreed to comply with the state's wish, but insisted on doing it in a different way. The town council decreed that motorists could drive at any speed they chose—except 71 m.p.h., at which speed they would be prosecuted. When anyone was caught driving at 80 m.p.h. or 90 m.p.h., it was argued that in reaching such speeds he must have at some point moved at 71 m.p.h. Therefore he was guilty. The space-twin is in a similar situation. It is not his speed

* The famous cosmologist Fred Hoyle was once interrupted at a lecture by a heckler who demanded, 'Professor, tell us, what *is* mass? What *is* gravity?' For once in his career, Hoyle was nonplussed. 'Well,' he said, I can't tell you what they *are*, but I can tell you what they *do*.' 'What they do !' jeered the heckler. 'That's not science, that's technology!'

through the cosmos which causes him to age more slowly; it is the *acceleration* which he has undergone to reach that speed. He has accelerated against the inertia which otherwise would have kept him flying in some perpetual orbit where his engines would not be needed at all. Inertia was the phenomenon incorporated by Newton into his First Law, which states that 'every body or object remains in the same state of rest or of uniform motion unless acted upon by a force.' And so inertia behaves in exactly the same way as gravity.

Einstein pounced on this point. He made the staggering assertion—which Newton would have considered mad—that inertia and gravity behave in the same way because they are two different words meaning exactly the same thing! Consider a heavy cannonball and a ball made of light wood. Contrary to Aristotle's belief, they fall to the ground at precisely the same speed.* The cannonball would prefer to rush to the ground much faster than the wooden ball. But while gravity pushes it down, inertia holds it up. And so it falls at a limited speed. The wooden ball, because it is lighter, would prefer to fall more slowly. But it in turn is subjected to much weaker inertial forces, and so it is compelled to move at the same speed as the cannonball. Ignoring the effects of air, a body with the mass of a mountain would fall no faster and no more slowly than a feather. Gravity and inertia always cancel out each other. Gravitational mass is always exactly proportional to inertial mass.

The balls only hit the ground because they are released at a low altitude. If an object was released at an altitude of about 300 miles, it would never hit the ground, although it would continue to 'fall' for the rest of eternity. It would be a satellite in orbit. I make this rather obvious point to show that objects falling to the ground and objects flying in perpetual orbit are doing exactly the same thing; they are following the line of least resistance through the Universe. No instruments exist sensitive enough to detect any curvature in the fall of the ball that hits the ground, but anyone can see that the satellite is orbiting in a curved path just as the Earth itself follows a curved path round the Sun. Every object in

---

* As demonstrated on the surface of the Moon by the astronauts of Apollo 15 in August 1971. A feather and geological hammer were dropped together. They hit the ground at exactly the same time.

the known Universe is in some kind of curved orbit around some other object. Newton described gravity, which creates these orbital paths, as a 'force' which 'pushed' and 'pulled' objects through their orbital motions. But Einstein, together with his Polish colleague Hermann Minkowski, gave the completely different explanation which is accepted by most experts today. To do this, they had to invent an entirely new conception, called 'four-dimensional space-time'.

Some people regard the fourth dimension as a sort of mystical entity having something to do with ghosts: it is nothing of the kind. It is the dimension of time or duration, without which no material object can exist. Every object has four dimensions. Consider a brick; it has length, width and height, three dimensions at right angles to each other. Time must be viewed as a fourth dimension at right angles to the other three. We might give the dimensions of the brick by saying, 'It is four inches by six inches by nine inches by a hundred years,' and mean that at the end of a hundred years it will have crumbled into dust and ceased to exist. We have seen how a spaceship's four dimensions of space and time change during high speed. Relativity physics insists that time is a dimension which behaves in a similar way to the other three.

Rejecting Newton's theory of gravitation, Einstein and Minkowski showed why the Earth follows its elliptical orbit round the Sun. The Sun does not 'pull' on the Earth as Newton believed. Instead, *its large mass causes space and time to curve or warp in the area surrounding it.* The Sun's mass creates a corridor whose walls are made of curved space and curved time, and the Earth travels eternally through this corridor. The idea that time also is curved might sound terrifying and incomprehensible. But this is only because we have never encountered anything of the kind during our Earth-bound lives. Every star and galaxy in the Universe is busy at this game of creating gravitational fields that result in corridors consisting of curved space-time. There are an infinite number of such corridors—and that is the essence of the General Theory of Relativity.

The General Theory has invalidated most of the geometry which we learned at school. Euclid's laws about parallel lines that never meet, about triangles that must possess 180 degrees and no more, are, in absolute terms, false, because they assume that the shortest route between two points is a straight line. Einstein and Minkowski showed that in reality there is no such thing as a straight line. All lines, if extended long enough through space, are curved. A beam of light which sets out on a journey through the Universe will travel in a complete circle and return to the point where its journey began. This explains Einstein's famous joke to reporters that if a man with phenomenal eyesight looked up into the sky, he would see the back of his own head. He would, of course, have to wait billions of years before the light image of his own head travelled round the Universe, but the joke helps to explain that the light image takes a curved route because space is itself curved.

The General Theory was first proved in 1919 during an intermission in a tropical thunderstorm in the lonely Portuguese colony of Principe, in the Gulf of Guinea in West Africa. Arthur Eddington, a talented young astronomer who had followed Einstein's work with excitement, led an expedition there to photograph stars during a local eclipse of the Sun. He wanted to find out whether the light beams from the stars were deflected from their courses by the Sun's mass, as the General Theory predicted.

The eclipse at Principe was due to take place at 2.15 p.m. local time, and it would last only five minutes. Rain poured down throughout the morning as Eddington was trying to prepare his telescope and camera. He tried to protect the apparatus inside a tent, but rain constantly seeped in, threatening to ruin the film. The sky partly cleared just before the Moon drifted in front of the Sun. He was able to take sixteen photographs of a star near the Moon's rim, but the scudding clouds made him doubtful whether they would come out. Fifteen of the shots were blurred and useless, but the sixteenth was a fine shot showing a cluster of bright stars. They were carefully examined to see if their positions were different from normal.

Einstein was waiting in suspense to hear the results. He had gone to Leyden University in Holland to lecture and to hear one of the famous talks of Hendrick Lorentz, the Dutch theorist who had written one of the essential equations of the Special Theory. Einstein was lecturing when Lorentz appeared on the platform waving a telegram. It was from Eddington in London. It announced that the star positions in the Principe photograph had changed by an average of 1·7 seconds of arc, well within the range Einstein had predicted. Einstein was unable to speak. There was a prolonged burst of cheering from the hundreds of assembled students and professors.[5]

The news astonished and delighted a public that was wearied by the miseries of post-war Europe. Einstein was besieged by reporters, amateur astronomers and politicians. He was even lobbied by a cigar manufacturer who wanted to market 'Einstein cigars'. They grew shorter as they were smoked; perhaps they were contracting in the direction of motion. Men rushed out at night with torches and shone them over the surrounding fields. Many swore that they could see the light beams being bent downwards by the Earth's gravity. A cartoonist depicted two detectives in Sherlock Holmes deer-stalkers with torches whose beams turned a right angle round the corner of a house to alight on a burglar. 'Elementary, my dear Einstein,' said one to the other. There were, of course, many people who were contemptuous of the new theory. Some professional astronomers, even some of those who had accepted the result of the Michelson-Morley experiment, dismissed the whole thing as utter nonsense. One physicist, Josiah Gibbs, remarked, 'A mathematician may say anything he pleases, but a physicist must be at least partially sane.' Emile Picard, permanent secretary of the French Academy of Sciences, stated, 'On the subject of Relativity I see red.' Clergymen said the theory was blasphemous. Sailors who had long been using the stars for navigation and who had always found them in the same places declared that Einstein was insane.

It required, above all, a tremendous intellectual effort to reject the old ideas and accept Einstein's strange new concepts. Laymen sometimes found this almost as difficult as did elderly scientists. A 1928 editorial in the *New York Times* said:

Tennyson claimed for faith the function of believing what we cannot prove. The new physics comes perilously close to proving what most of us cannot believe ... Countless textbooks on Relativity have made a brave try at explaining it, and have succeeded at most in conveying a vague sense of analogy or metaphor, dimly perceptible while one follows the argument painfully word by word, and lost when one lifts one's mind from the text.[6]

Einstein's inexorable logic, which left no room for loop-holes, in time convinced the most sceptical. Today, among mathematicians, physicists, astronomers and cosmologists, there is almot unanimous acceptance of the two theories. The experiment of Michelson and Morley, the experiment of Eddington, have been carried out again and again with no difference from the first results. The relationship between mass and energy, which appeared in the Special Theory as $E = mc^2$,* has given us nuclear energy and nuclear bombs. In the words of Professor John A. Wheeler of Princeton, 'No purported inconsistency with the General Theory's predictions has ever stood the test of time. No logical inconsistency in its foundations has ever been detected. No acceptable alternative has ever been put forward of comparative simplicity and scope.'[7]

Wheeler has today given the General Theory a staggering new twist. It will be shown in the next chapter how his interpretation of it, and his creation from it of a radical new cosmology, is beginning to reveal to us the existence of those hidden paths in space which might one day enable a spaceship to travel from Earth to a habitable planet in another solar system in the same amount of time as it now takes to fly from London to Sydney.

* This all-important equation (usually computed in metric units) means that energy in ergs equals mass in grams multiplied by the square of the speed of light (c^2) in centimetres per second. In other words, a tiny amount of mass can produce a gigantic amount of energy.

9

Voyage Through a Hidden Universe

Science fiction on the whole ignores the theories of Relativity. There are for instance novels which deal with the political intrigues of galactic empires, in which an emperor has established his court on some planet near the centre of the Galaxy, and from which he must rule by sending and receiving missions and messages from remote provinces on its periphery. The authors try in various ways to evade the ruling of the Special Theory that a message travelling at the speed of light would take fifty thousand years to move from the galactic centre to the periphery.* It would take another fifty thousand years to return—by which time the emperor and all his works would have passed away. These difficulties are usually ignored, and we are told simply that the emperor sends a message and receives a reply to it on the following day. Even more remarkable are the journeys through the Galaxy of the imperial agents. One novel contains the curious line, 'We are outpacing light, and from the portholes we could see the stars whizzing by.' This is about as intelligent as the line from the landlubber author of sea-stories: 'The great sails filled, and the ship sped at full speed across the flat-calm sea.' For, apart from other difficulties, if the astronaut were outpacing light, he could never see the stars 'whizzing by'. It is doubtful if he could see anything at all. He would be going so fast that the light from the stars would never be able to catch up with him.

One novelist, however, describes interstellar journeys which never strain our credulity. Dr Isaac Asimov, himself a scientist, has sold millions of copies of his galactic empire novels while the works of his rivals have been remaindered. This is perhaps because Asimov knows his Relativity, and has the instinctive *feel* for

* Because our Galaxy is about 100,000 light-years in diameter. A light-year, the distance which light travels through space in one year, is a common unit of distance in astronomy. It is about six million million miles.

guessing how the faster-than-light problem may eventually be overcome. Consider this passage from his thriller, *The Stars Like Dust*, which sounds almost prophetic. The travellers are departing from Earth in their space-liner on their way to a distant star. The following announcement is made:

'This is the captain speaking. We are ready for our first jump. We will be temporarily leaving the space-time fabric to enter the little-known realm of Hyperspace, where time and distance have no meaning. It is like travelling across a narrow isthmus from one ocean to another, rather than circling a continent to accomplish the same distance. There will only be minor discomfort. Please remain calm . . .'

It was like a bump which joggled the deep inside of a man's bones. In a fraction of a second the star view from the portholes had changed radically. The centre of the great Galaxy was closer now, and the stars appeared to thicken in number. The ship had moved a hundred light-years closer to them.[1]

Asimov did not know it when he wrote this passage, but a large number of cosmologists today are convinced of the existence of Hyperspace, that strange-sounding region that is like another universe alongside, or *within* our own. In science it is called Superspace, and the word came into parlance soon after being discussed in an important paper[2] co-authored in 1962 by Professor Wheeler, a co-inventor of the hydrogen bomb, who has perhaps brooded more deeply than most mathematical physicists on the hidden meanings of the ten field equations of the General Theory of Relativity. Our knowledge of Superspace is limited, but it appears to have some definite characteristics. It is a huge region, permeating every part of the known and visible Universe. Entrances to it and exits from it are believed to exist everywhere, in the spaces between the galaxies, in the spaces between the stars, and even on the fringes of our own Solar System. Wheeler suggests that the true shape of the Universe may be that of a doughnut—or of any object whose general shape is that of a solid ring. All the stars and galaxies that we can see are on the curved, solid part of the doughnut, while the hole in the middle represents the mysterious region of Superspace. A signal, or a spaceship, travelling the conventional routes across the doughnut's curved surface, takes a long time to complete its journey because of the huge size of the Universe. But in the inner hole of Superspace, as we shall see, the journey would be much more rapid since there the ordinary laws of physics are altered.

We have strong evidence for the existence of Superspace. These pieces of evidence are doubly satisfactory since they hold good for either of the two main theories of the history of the Universe. The 'steady state' theory, long favoured by Fred Hoyle, which holds that the Universe as a whole is eternal, and that while stars and galaxies may die, there have always been and always will be other stars and galaxies to take their places, is satisfying on several counts; but it does not work unless we accept the existence of Superspace. This theory, which Hoyle has vigorously propagated—although it was originally proposed by Hermann Bondi and Thomas Gold—describes a Universe in a state of 'continuous creation'. Stars die and others take their places, and this process continues for eternity. An enormous amount of hydrogen is available in galaxies for the continuous formation of new stars, but not enough. It is true that Hoyle and his colleagues require only a very low rate of increase of hydrogen, but small though these continuous replenishments of hydrogen may be the hydrogen must come literally from nowhere.[3] And since it is impossible for anything to come from nowhere, we can only reason that it comes from Superspace.

It has been lately realized with something of a shock that the other main theory of the history of the Universe, the 'Big Bang' or expanding Universe theory, does not work either without Superspace. This well-known theory proposes that about 12 billion years ago the entire Universe was contracted into a single incredibly dense 'primeval atom', which was about the size of our present Solar system. The late George Gamow, the Russian-born physicist who helped to propound this spectacular theory, named this object *ylem* (pronounced 'eelem', an old Greek word for primordial matter). Suddenly, according to Gamow and the many cosmologists who today accept this theory, *ylem* blew up in the most shattering explosion there has ever been. It must have been like a supernova, which is the total explosion and disintegration of a star, but billions and billions of times more violent. Great chunks of matter blasted off in every direction, some of them at almost the speed of light.[4] The Universe is still expanding, some 12 billion years later. The further into space we look, the faster the galaxies are receding from us. But astronomers at the California Institute of Technology, home of the supremely powerful Mount Palomar telescope, have calculated that the expansion of the Universe is beginning to slow down, and that in about 70 billion years' time it will stop expanding; the galaxies will then start rushing together again, perhaps to form another *ylem.*

But the mystery is *why* the expansion is beginning to slow down. Large masses in the Universe have obviously created a gravitational field that is slowing down the distant galaxies. But what large masses? The inescapable fact is that all the matter of the known Universe contains less than one-tenth of the mass sufficient to put a brake on the expansion. As Wheeler declared, in true detective-story style, this is the 'case of the missing matter'.[5] The suggestion that the missing matter may be located in huge clouds of gas that float between the galaxies has been convincingly refuted;[6] the required amounts of mass simply are not there. Wheeler and others have drawn the only possible conclusion; the missing matter is located in Superspace.

Superspace is indispensable to any cosmology that takes all the observed facts into account. A Universe without the extra arena of Superspace would be like an actor without a stage. Superspace, in Wheeler's cosmology, is eternal. It is the everlasting physical background to the *ylem*-to-*ylem* cycle of our own Universe and of all other universes, past and future. Whatever universes succeed this one, or came before it, they are always surrounded and permeated by Superspace.

The nature of Superspace is a very difficult subject. Even thinking about it seems like trying to swim through a sea of mud. 'It is like chasing after Merlin,' says Wheeler. 'One moment it is a rabbit, and the next a gazelle. And just as you reach out to touch it, it turns into a fox, or a brightly-coloured bird fluttering on your shoulder. It is the place where smoke comes out of the computer because all the classical laws of space-time break down.'[7] It is the region, he goes on to explain, into which all the planets and stars and galaxies will eventually disappear, swirling down a time tunnel in space, like water gurgling out of a bath.

In it, time stands still, and the events of a billion years are compressed into a split second. Descriptions of Superspace may sound almost mystical, but our awareness of its existence is surely the greatest post-war advance in fundamental physics. Superspace, the super-Universe, solves all manner of cosmological problems. Every theory of our Universe which excludes it sooner or later meets the same difficulty; events that require much more available matter than conventional astronomy provides for. A theory which excluded Superspace would have to be one in which light travelled everywhere in straight lines. It would have to reject Relativity in its entirety, which would mean denying the existence of nuclear energy. And so such theories cannot be of much value.

Until 1970, the existence of Superspace was wholly theoretical. The argument for it seemed irrefutable, but no one had found a way to prove its existence by experiment. Then in February 1970 an extraordinary announcement was made. The journal *Physical Review Letters* published a report by Joseph Weber, of the Princeton Institute for Advanced Study, under the modest heading 'Gravitational Radiation Experiments'. Weber stated that two synchronized instruments, 600 miles apart, one in Maryland and one in Chicago, had detected waves of gravitational energy sweeping in from the denser parts of the Galaxy.[8] These were not steady pulses of gravity, such as one might expect from large bodies of matter. They came instead in violent bursts, coming apparently from nowhere and never repeated. Weber reported that these bursts came at least once a day from different parts of the Galaxy. His paper, which he soon followed by another later in the year,[9] astonished cosmological scientists. Only events of tremendous violence could be producing such waves. It was concluded that entire stars were being obliterated, vanishing completely from the Universe into what are termed 'black holes'.

When a star has burned up all its fuel by converting all its hydrogen into helium, it either swells into a red giant or explodes. It then begins to collapse; the outer parts begin to fall towards the centre, at first slowly but then more and more rapidly. It can end in three ways, depending on the original mass of the star. An average-sized star like the Sun will collapse into what is called a white dwarf, a body about the size of the Earth. It would be extremely dense, and one piece of material the size of a sugar cube from a white dwarf would weigh about five tons. But there the collapse would end, and the star would remain in this state for billions of years. But a larger star, one that was originally 50 per cent more massive than the Sun, will not stop collapsing when it reaches the white dwarf stage. The gravitational forces are too strong to be checked by the repulsive forces of density. The atoms of the star actually destroy each other. Their charged particles collide and neutralize each other. All that remains are the neutral particles of the crushed star, giving the name *neutron star* to the fantastic object that remains. This body, still with a mass 50 per cent greater than the Sun's, is now no more than about 10 miles in diameter. One cubic inch of it would weigh 100 million tons. Its gravity is so strong that light can hardly escape from it. But neutron stars are quite easy to find, because they rotate very rapidly, giving off radio pulses which have given them the nickname of 'pulsars'. Pulses from the famous neutron star in the Crab Nebula, some four thousand light-years from the Earth, show that it is rotating no less than 30 times per second.*

An even stranger fate awaits a collapsed star which was originally twice the mass of the Sun. Here, the gravity is so strong that not even the neutron particles can resist it. Gravity overpowers all repulsive forces until every bit of matter of the star is annihilated. The star has completely vanished, apparently into nowhere. It has gone into a *black hole*, and nothing of it remains in our Universe except that single burst of pure gravitational energy which Weber has detected. When the existence of black holes was first suggested in 1939 by the nuclear physicist J. Robert Oppenheimer and his student Hartland Snyder, it was believed that they were stars which had only partly collapsed. They were called 'black', or invisible, because their gravity was thought to be too strong to permit any light rays to escape. The crushed star was still there, yet it was invisible. But a genuine black hole, it is now believed, is not invisible because light cannot escape from it; it is invisible because it is not there any more. A body of matter,

---

* When this pulsar was first discovered in 1967 by the Cambridge astronomer Anthony Hewish and his assistant Jocelyn Bell, it was thought that intelligent signals from another race had been detected, and the discovery was kept secret for several weeks.

nearly a million times more massive than the Earth, has simply vanished.

Weber has given us an impossible paradox. If these events are occurring almost daily, as his results show, then the Universe cannot possibly continue for anything like the 150 billion years predicted for its future. Our own Galaxy, for instance, contains 100 billion stars and is known to be 10 billion years old. But if one star is vanishing from it each day, and if we assume that these disappearances have been continuing at a similar rate for billions of years in the past, then the Galaxy could not have existed for more than 270 million years, about one fortieth of its actual known history. Apart from any other consideration, if the Earth was only 270 million years old (instead of its true age of 4·5 billion), then advanced life on Earth would be out of the question. There would not have been nearly enough time for dinosaurs to have evolved, let alone people. Moreover, observations of the orbits of stars rule out any possibility that the Galaxy is sustaining any net loss of mass on anything like the scale which Weber's results would suggest.[10]

But this is not to say that Weber's results were wrong. It might be useful to visualize the riddle in industrial terms. Imagine a factory which employs 100 men. For as long as we can remember, it has always employed 100 men, never more or less. But every day one of the men is dismissed. And yet the work force mysteriously remains constant. Instead of being empty after 100 days, the factory still employs 100 men. The only possible explanation is that the dismissed men are being reinstated in a process which we cannot yet observe.

So it seems to be in the case of our Galaxy of stars. A superb solution on these lines was given a year later by Robert Hjellming of Green Bank, West Virginia, in a letter to *Nature Physical Science*.[11] If the total amount of matter in the Galaxy is to remain roughly the same for extremely long periods, Hjellming suggested —and all physical laws demand that it must—then matter must *appear* as fast as it disappears. Stars would re-emerge into our Universe through a 'white hole' soon after they disappeared through a black hole. The process of collapse would occur in reverse. The phrase 'anti-collapse' has been coined by one scientist to describe the re-emergence into our own Universe of a star which previously vanished at another point in space.[12] In short, says Hjellming, stars must disappear in one place and reappear in another. To make this journey, the star will have travelled through Superspace. Our Universe is therefore 'multiply connected' by black holes and white holes, just as London is 'multiply connected' by its innumerable underground railway tunnels. Hjellming's theory solves also the 'case of the missing matter', upon which all previous models of the Universe have foundered because of difficulties with mass-conservation laws, which state that matter cannot be created or destroyed. 'In each Universe', writes Hjellming, 'conservation laws may not apply whenever the volume considered contains a "hole"; however, they are strictly obeyed when both universes are considered.'

It may be wondered what all this has to do with the 'four-dimensional space-time' of Einstein and Minkowski. Space-time is the clue to it. It is the justification for Superspace. If space-time consists of curved corridors, then it must be assumed that the walls of such corridors consist of some kind of tangible fabric. This is the answer to the riddle of how empty space provides a medium for gravity; space is not empty at all. It is filled with particles consisting of pure gravitational energy (Weber has detected unusually strong bursts of them) which John A. Wheeler calls 'geons'. Geons, massed together, make up the solid structure of the curved walls of space-time, and it is through this solid, sponge-like structure that holes exist. With our present limited powers of detection, we might assert that geons do not exist because we cannot see them or touch them, and because they seem to be composed of absolutely nothing. But anyone who believes that should try jumping out of a window, and he will find that geons will employ tremendous energy in hurling him downwards to the ground. Particles that behave like this cannot be dismissed as non-existent. The Special Theory tells us, in the equation $E = mc^2$, that energy is directly attributable to mass. Energy without mass is an impossibility, and this alone gives us grounds for suspecting that the geons that comprise the fabric of space-time do, in fact, consist of solid matter.

'Do geons exist, or do they not?' Wheeler asks. 'Do they have mass or do they not?' It is easy to follow his triumphant proof that the answer to both questions is yes. It is undeniable that the photon light particle possesses both energy and mass, as anyone who has been blinded by an oncoming car headlight can testify. The photon obeys the gravitational laws to which mass is subject; we have seen how photons curve round a massive object like the Sun. There is no difference in this respect between the behaviour of photons and that of geons. The only thing that has confused us is that photons are visible and geons are not. For this reason only, it has taken much longer to recognize the nature and the existence of geons.

A geon [writes Wheeler], though built of curved empty space and completely free of any so-called 'real mass', is nevertheless endowed—by reason of its radiation content—with mass of its own. This mass holds it together against leakage losses for a time very long in comparison with the travel time of radiation through its interior. The geon moves through space as a unit. It responds to the gravitational fields of other masses. It also exerts forces on them. It provides a completely geometrical model for mass.[13]

And so, when bunched together, geons make a wall of solid matter. Their nature and behaviour are the basis of modern 'geometro-dynamics', the strange-sounding science which Wheeler and others have developed from the General Theory, and which means the geometry of curved 'empty' space, or the dynamics of geometry itself. In Wheeler's words, the long-neglected General Theory is today 'spawning new riches, possessing properties far more powerful than even Einstein foresaw. To ignore because it seems too far removed from reality might be like having ignored $E = mc^2$ back in 1939.'[14]

Wheeler envisages 'worm-holes' in the structure of space-time, similar in their dynamics to black holes and white holes, but on a very much smaller scale. Worm-holes are to be found everywhere in space-time. Wheeler draws the analogy:

Space is like an ocean which looks flat to the aviator who flies above it, but which is a tossing turmoil to the hapless butterfly which falls upon it. Regarded more and more closely, it shows more and more agitation, until . . . the entire structure is permeated everywhere with worm-holes. Geometrodynamic law forces on all space this foam-like character.[15]

The solid, curved walls of space-time are flawed by these tiny holes, just as the structure of almost any solid surface would be seen to contain holes if sufficiently powerful microscopes existed. Worm-holes are the entrances to Superspace and the exits from it. A signal, or conceivably a spaceship, could pass through a worm-hole, enter Superspace and then emerge from Superspace through another worm-hole into another part of our own Universe. But what purpose would there be in undertaking such a mad and probably dangerous voyage? The answer is that this will solve our faster-than-light problem. For inside Superspace, *time does not exist*, and a journey through any part of it would therefore be instantaneous by whatever clock the voyage-time was measured. This is logical since a traveller who has entered it and departed from space-time has left time behind him. Every event in Superspace occurs simultaneously; this region has neither past nor future, but an eternal present.

Wheeler and his colleagues are emphatic on this point. 'In Superspace,' he said, 'the question, "what happens next?" is devoid of content. The very words, "before", "after", and "next" have lost all meaning. Use of the word "time" in any normal sense is completely out of the question.'[16] The analogy of the brick may help us; normally, it has four dimensions, those of length, width, height and time. But remove the time dimension, and the brick must instantly cease to exist in any arena where we can measure it or detect its presence. The traveller who enters Superspace will disappear from our field of vision just as suddenly. But to the imaginary people inhabiting the distant planet that was his destination, he would suddenly appear, apparently from nowhere, but in reality from the hidden regions of Superspace.

The interior of Superspace remains mysterious. It has two characteristics that seem contradictory. An object can move in and out of it in no time at all, and yet it contains over ten times more mass than the visible Universe. This would suggest that massive bodies remain 'trapped' in Superspace, themselves experiencing no time duration at all, but as far as we are concerned remaining there for as long as the Universe endures. Alternatively, Superspace may be like a busy street, which is always filled with people in transit, but who pass through quickly on their way to somewhere else. But whether the missing matter is trapped or in transit, it is unlikely to impede those objects that want to use Superspace as a short-cut.

The crucial 1962 paper in the *Physical Review*, co-authored by Wheeler and Robert W. Fuller, which first introduced these ideas, was entitled 'Causality and Multiply Connected Space-Time'. This meant that space was 'multiply connected' by worm-holes, just as in Hjellming's later theory. 'One may ask', they wrote, 'if a signal travelling at the speed of light along one route could be *outpaced* by a signal which has travelled *a much shorter path* through a worm-hole.' While cautiously emphasizing that their entire interpretation of geometrodynamics was so far little more than a mathematical exercise, and for all they knew might not apply in the real Universe, they concluded that Einstein's and Minkowski's conception of curved space-time allowed such worm holes to exist, and such fantastic journeys through them to be made.[17] Since then, Wheeler and other cosmologists have become more positive that their interpretation was correct.

It is uncertain whether such an interstellar journey would involve one single journey through Superspace, moving in a straight line across the hole in the middle of the doughnut, or whether it would be a *series* of short-cuts, similar to Asimov's idea of a series of 'jumps' through 'Hyperspace'. It is easy to visualize such a jump. Take a sheet of paper and mark Earth as a dot. Mark another dot at the bottom of the paper to represent a point in space five light-years away. Travelling directly down the paper, a signal would take five years to make the journey. But fold the paper backwards until the two dots touch each other. Now we have a very much shorter path. Complete the experiment by cutting out the dots to make two little holes in the paper. These are the worm-holes through which the signal has leaped into and out of Superspace, taking no time at all to make the journey. Another jump could then be made, through another part of Superspace, and so on until the final destination was reached.

Other scientists have independently reached similar conclusions to those of Weber, Wheeler and Hjellming. After a star has disappeared through a black hole, its matter, according to Kip Thorne of Caltech, 'might then emerge, bubbling upwards like a spring in the mountains, in some other region of our own Universe'.[18] There might be anxiety that a spaceship that made the same journey would itself be crushed by gravitational forces. But some people believe that this need not happen. A star in some circumstances, or a spaceship expertly navigated, could emerge *undamaged* at another point in space. Margaret Burbidge, director of the Royal Greenwich Observatory, Igor D. Novikov, of the Institute of Applied Mathematics in Moscow, and James M. Bardeen, of the University of Washington, have all produced models of a stellar collapse in which precisely this happens.[19]

Without discussing their extremely intricate mathematics, I will say only that some of these models involve a star or other matter vanishing into a black hole without going through the crushing process of total gravitational collapse. As the cosmologists Stephen Hawking and G. F. R. Ellis explain in their monumental 1973 work, *The Large Scale Structure of Space-Time*, an interstellar dust-cloud could, without being crushed, 'pass through a worm-hole into another region in space-time.'[20] It is important to note that most of these models were worked out before Weber published his paradox and Hjellming produced his solution. The Israeli scientist Yuval Ne'eman, co-discoverer of the famous Omega-minus particle, was also unaware of their conclusions when he drew an analogy which has attracted a certain amount of ribald comment. He saw the Universe as comprising two 'trouser-legs', the seen and the unseen, with 'tunnels' connecting them.[21] Thomas Gold has proposed a system of multiple universes, each joined on to the next.[22]

This kind of thinking is encouraged by the equations of the General Theory. Einstein himself, in a long-forgotten paper of 1935, hinted at the existence of 'bridges' connecting two or more widely-separated parts of the Universe.[23] 'It is quite possible', Einstein remarked on another occasion, 'that other universes exist independently of our own.'[24] Some people, in exasperation at these mad-sounding notions, have urged that the General Theory be abandoned and replaced with something less bizarre. But Roger Penrose, of London University, has silenced these somewhat emotional protests by pointing out that similar phenomena will follow from any other viable theory of space-time. The human applications of geometrodynamics are still at the completely theoretical stage that nuclear energy was in before

Rutherford and Cockcroft split the atom in 1932. A huge amount of mathematical work has to be done before any signal can actually be transmitted through Superspace. But a manned landing on the planet Mars, which will almost certainly take place before the end of this century, may prove an ideal opportunity for an experiment. Mars at its closest approach is about three light-minutes from Earth. Light rays or radio signals normally make the journey in three minutes. By the time that the manned Mars mission is ready to leave, scientists and engineers may have been able to devise a 'Superspace transmitter' to be established at the Martian base, which would send signals back to Earth *in less than three minutes*. The prospect of trying to construct such a device appears as difficult and as complicated to us today as a manned Moon landing seemed in 1950. Now, as then, the technical problems seem at first sight overwhelming. It will no doubt be necessary to design a computer of scarcely imaginable sophistication to do the required mathematics. A hole in space-time will have to be found, or perhaps 'made'. And then a signal must be induced to enter it—we do not know which will be more suitable, a radio or an optical signal. And, having entered Superspace, the signal must leave it through another hole in the neighbourhood of Earth. We do not yet know how to make this happen. Even a partial success would be a magnificent achievement.

If the signal vanished completely, we would know that it had entered Superspace. It might have emerged at some point far from Earth, or it might have been trapped inside; there would be no means of knowing. The experiment may have to be carried out many times before there is a total success, and the accelerated signal reaches Earth. And what a success that would be! It would be more exciting than the results of all the elaborate searches for primitive Martian plant life. It would mean that man was liberated from the single Solar System in which he has so long believed himself to be a prisoner. For if a signal could make this faster-than-light trip, it would no longer seem impossible to design a spaceship that would do the same. It may not be improper to illustrate this idea with the line from Shakespeare's *Henry VI Part 3*—'Where the fox hath once got in his nose, he'll soon find means to make the body follow.'

These advanced ideas give substance to Haldane's view that the Universe is 'queerer than we can imagine'. Wheeler, Hjellming and Weber present us with a cosmos far richer in complexity and mystery than anything which the conventional astronomers have dared to propose. We unlock a door, only to find another locked door concealed behind it. Behind that, too, is found yet another door, a process which perhaps will continue for as long as the human mind retains its power of inquiry and hypothesis. Even today, some scientists occasionally announce that we know as much about the Universe as we ever shall. The advanced cosmologists turn such statements into nonsense almost as soon as they are published. Modern cosmological knowledge, having been given its first impetus in 1905 by the genius of Albert Einstein, has surely centuries to go before it becomes all-embracing. It is not yet a century old; one might be pardoned for imagining that almost everything important remains to be discovered. It is very difficult in these conditions to take seriously Edward Purcell's conclusion that manned galactic exploration 'belongs on the cereal box'.

We nevertheless have to face the possibility that Superspace may not prove navigable. It is possible that some of the cosmologists have been wrong in proposing that an object can traverse Superspace without being crushed by gravitational forces. And even if they were right, it may prove impossible, for some reason or other, for spaceships to imitate this feat. This latter possibility is much less likely, for if the thing *is* possible, then men will surely find a way to achieve it. Yet there still remains a small chance that all modern interpretations of the General Theory are wrong, or that Hjellming was wrong in proposing the existence of 'white holes', or that Weber wrongly interpreted his signals. Should we therefore expect that failure of the General Theory would destroy all hopes of interstellar travel?

Certainly we should not, since there is another method of travelling rapidly to the stars that does not depend on Superspace or holes. This is by exploiting the Special Theory, particularly that equation of it which states that a moving object ages more slowly than a stationary one. As I explained earlier, it would be possible for a man to make a round trip lasting a few months to the nearest star, Proxima Centauri, and find that the Earth had aged centuries during his absence.* An experiment in 1971 proved that a human being's ageing processes slow down at high speed just like those of any other object. Two American physicists, Joseph Hafele and Richard Keating, boarded a jumbo jet in Washington, D.C. They carried with them an atomic clock that recorded time with an accuracy of billionths of a second. Another clock, precisely synchronized with the moving one, was left behind in Washington. Returning from their round-the-world-trip, during which they flew at 600 m.p.h., they found that the airborne clock was about 90-billionths of a second behind the stationary clock in Washington.[25] Six hundred m.p.h. is of course less than a millionth of the speed of light, and so the difference between the airborne and stationary clocks was correspondingly small.

But imagine a much faster journey, a round trip to Proxima Centauri. If the astronauts accelerate on each leg of the journey to 99·999 per cent of the speed of light and, taking a dislike to Proxima Centauri when they see it, immediately set out for home they will return after only ageing about three weeks. But the Earth will have aged nearly nine years in their absence.

By this means, if the General Theory and geometrodynamics fail us, we have a sure means of reaching the stars. It will be a much less satisfactory method than through Superspace, since Earth-men will have to wait ages for their astronauts to return looking irritatingly fresh and spruce, but it will work. The problem of *how* to build engines that will accelerate a starship to nearly 670 million m.p.h. I will leave to the technologists. It is a matter of devising an engine of such efficiency that while occupying very little space it will keep up a steady acceleration for about 100 days, and which does not, for the sake of comfort, exceed one G force. Some idea of the power of such an engine may be given by comparison with the third-stage engines that propelled the Apollo spaceships away from Earth orbit towards the Moon. The Apollo engines were fired for three minutes; our engine will need to run continuously for 100 days. This is obviously far beyond our technical capacity today. Mountains of literature have been published on the feasibility of such engines, and I will not

* The reader will find, in Appendix 2, a more detailed account of the extraordinary effects of the Special Theory.

summarize them here.[26] It need only be said that, judging by past technological performance, the problem will be solved if a solution is indeed required. Present evidence suggests, however, that Superspace will remove all need for engines of such colossal power.

Whatever kinds of interstellar societies are created, they will be immune to that particular disease that now afflicts man, that of tension caused by overcrowding. There can hardly be any lack of room in a Galaxy that contains at least 100 billion suns, of which there are perhaps 100 million with one or more habitable planets. Even if the resources of all these become exhausted, at some future date too remote to be imagined, there will be billions of other galaxies to be explored, each with their hundreds of billions of stars. While the Universe seems to be finite in size, for all practical purposes the number of possible worlds may be considered infinite. The Church made this suggestion seven hundred years ago. The Bishop of Paris, in 1227, ruled that it is wrong to deny God's power to create as many worlds as He pleases.

10

The Search for Habitable Worlds

Questions about the number and nature of worlds in the Universe have fascinated people in all ages. Plutarch relates how Alexander the Great once flew into a tantrum because his friend Anaxarchus told him that there was an infinite number of worlds. Alexander exclaimed, 'Do you not think it a matter of lamentation that, when there is such a vast multitude of them, we have not yet conquered one?'[1] Giordano Bruno, one of the most aggressive philosophers of the Renaissance, challenged established opinion of his time by preaching that every star was a sun, and that every sun had planets. Even more intolerable than these heretical views was the fact that his lectures attracted crowds of students, while his colleagues mumbled away in empty halls. In 1592 he was denounced to the Inquisition in Venice. In the words of one scientific historian, no defendant since Socrates worked harder to secure his own conviction.[2] He jeered at his trial judges, boasting to their faces that they were frightened of him. He was burned alive at the stake, the first and only martyr to the theory of the plurality of worlds.

The number of worlds *in the Universe* is likely to remain an academic question even for the thousands of future years whose events this book attempts to predict. Our own Milky Way Galaxy, although only one galaxy among tens of billions of others, is quite large enough, with its 100 billion suns, to occupy human explorers for an enormous period of time. A more practical question is, therefore, how many habitable worlds exist in our Galaxy? Aided by the instruments which Bruno lacked, we can make an informed guess of the approximate number. 'Habitable' in this sense must mean planets upon which human colonists can live in reasonable comfort. Conditions on the *majority* of planets are almost certainly intolerable to human life, as a glance at our own Solar System suggests. Of nine planets, which presumably are a typical sample since they orbit a typical star, only one is habitable for our kind of life. None of the others can be occupied without massive environmental alterations, of varying expense and complexity. A few calculations enable us to estimate the number of Galactic planets with environments sufficiently amiable for immediate settlement. Let us search for some.

The Sun is one of those average-sized stars which remain for at least 10 billion years on the 'main sequence' of the Hertzsprung-Russell Diagram. If the Sun was any larger, its life would be shorter. A star's period on the main sequence is the stable period of its life. This stable period begins when the star's internal temperatures have risen high enough for thermo-nuclear reactions to begin, and ends when the star's hydrogen fuel is used up, and the exhausted star becomes either a red giant or, if it is massive enough, a blazing nova.* As a general rule, the larger the star the shorter its life. The brilliant star Rigel, for instance, which is some 15 times more massive than the Sun and about 50,000 times more luminous, is unlikely to be much more than 100 million years old—less than a fortieth of the Sun's age. It would be idle to expect to find habitable planets orbiting a star so young. When our Sun was only 100 million years old, the Earth was certainly lifeless, with an unbreathable atmosphere of carbon dioxide, ammonia and methane. And since very massive stars tend to leave the main sequence at the age of about 150 million years, turning into red giants, novae or supernovae, it is almost inconceivable that they could have planets fit for life. A habitable planet must not only have a sun with a reasonable expectation of continued stability; it must also have plant life to sustain an atmosphere of breathable oxygen and nitrogen. Plant life appeared on Earth about 2 billion years ago, when the Earth was already between 2 and 2½ billion years old. Allowing therefore for about 1 billion years between the birth of the Sun and the formation of the Earth, we can say that any star much younger than 3 billion years will be useless to roving starfarers. It has been calculated from the Hertzsprung-Russell Diagram

---

* A star of very large mass can become violently unstable and explode into a nova at the end of its life. A catastrophic rise in temperature causes a partial stellar explosion, and the star gradually returns to normal. About 20 novae are seen in the Galaxy each year.

that a star cannot reach this age and still be on the main sequence if it has more than 1·43 times the mass of the Sun.

It may possibly be an error to treat the Diagram as if it were Holy Writ. We state, as if producing incontestable facts, that a star *cannot* have a habitable planet unless it has shone steadily for 3 billion years, and that it cannot shine steadily for this period if it has more than 1·43 Solar masses. It is possible, and perhaps even probable, that a Galaxy of 100 billion stars contains many exceptions to these two arbitrary rules. The first figure is based solely on our supposed knowledge of the Earth's history (which may be wrong anyway), and the second is the product of the statistical observation of very much less than 1 per cent of Galactic stars. There are doubtless thousands upon thousands of stars whose planets blossomed into life within 2 billion years of their own creation, just as there may well exist stars with two or three times the mass of the Sun, and which through some still incomprehensible process of slowed evolution, are still on the main sequence after 5 billion years. The Diagram, after all, represents nothing more than the sampling of a cross-section of the Galaxy. It is comparable with a public opinion poll which asks a small number of people how they will vote, and then multiplies the answers to predict the votes of a much greater number. But stars, fortunately, are much simpler organisms than people, and they do not tell us deliberate lies. We may therefore have much more confidence in the Diagram than we do in the average opinion poll. The search for habitable worlds must be guided by probabilities. If, for instance, the Diagram tells us that Star X is too young to have suitable planets, then the searchers will be well advised to ignore Star X. The Diagram could be wrong in this one case, but why bother? There will be many far more promising older stars.

Also to be avoided, in addition to the excessively large and young, are variable stars, which make a habit of blazing up brilliantly for a few days, and then returning to their normal luminosity. No planet within that star's ecosphere could survive such violent fluctuations. Such stars as Zeta Geminorum and Delta Cephei, whose brightness changes at regular intervals of days by several magnitudes, will have to be shunned by colonists.

Detailed three-dimensional charts of the Galaxy, showing the location of dangerous stars and stellar objects such as regular and irregular variables, periodic novae, super-massive stars, white dwarfs, neutron stars (pulsars) and black holes, will have to be kept in the libraries aboard all starships and in all interstellar admiralties. Only those stars approximating to the Sun's mass and age will be safe even for inspection from space at any distance closer than a few hundred million miles. According to some estimates, this 'list of wrong-size stars and of dangerous objects' will comprise as much as 70 per cent of the Galaxy, or about 70 billion stellar objects.[3]

The search is now much simplified. Knowing what kind of stars to seek and to avoid, let us turn attention to their planets. We have seen that a planet must have widespread plant life if it is to have an atmosphere that we can breathe. In general, only two kinds of planets will be of interest: Earth-type planets with abundant oceans, land masses and plant life, which will be ready for immediate settlement, and Venus-type planets, with atmospheres of rich carbon dioxide, which could be made ready for settlement after several years of algae treatment. Yet Earth-type planets, as we shall see, are likely to be so numerous in the Galaxy, probably numbering some hundreds of millions, that in the era of interstellar travel algae operations can be abandoned and Venus-type planets ignored. Carl Sagan's plan for Venus is important only because it will probably need to be carried out *before* the development of manned interstellar transport. In a Galaxy where worlds are plentiful, we can afford to choose the best. Our present confinement to a Solar System of extremely limited choice of action drives us to enterprises of great difficulty, which will be unnecessary in the almost boundless vastness of Galactic space. The search therefore can be restricted to Earth-type planets.

The habitability of the Earth, with its pleasant climates, its oceans, its forests, its mountains and its continents, is itself the result of something of a lucky chance, as the environmentalists constantly remind us. There is no rule in the Diagram that an average-sized star like the Sun *must* have a habitable planet. Consider how easily the Earth could have developed as an utterly different world if its mass, orbit and rotation had been only slightly altered at its creation. An understanding of the subtle effects of these facts will be essential to the searchers.[4]

Suppose that with everything else being the same, the Earth had twice its present mass, giving it a surface gravity 1·4 times greater.* The effect of the increased gravity would have been extraordinary. The evolution of marine life would have occurred much more slowly. The migration of sea-animals to the land would consequently have been delayed by millions of years. The whole evolution of life would have been correspondingly retarded. Mountains would be lower and broader in their bases. Trees also would be shorter with more massive trunks. Rivers would be deeper, since water, like all else, would be heavier. They would also flow much faster, and there would be less irrigation in the lands which they passed. Ocean waves would be lower, and the trajectories of their spray would be shorter. This would mean less evaporation and a much drier atmosphere. It is improbable, assuming that this heavier Earth was the same age as our own world, that life would have yet advanced beyond the stage of the great reptiles. But their numbers would be confined to the squatter and thicker-legged creatures. Animals with long necks and long legs, like giraffes, would have had no chance to evolve. Who can tell what form men would have taken, when they finally emerged some hundreds of millions of years from now? It seems certain that their physical appearance and their cultural patterns would have been utterly different.

Now suppose the opposite. Imagine instead the Earth with half of its present mass. This would give it a surface gravity 0·7 times less. Here we would have the very converse of the conditions prevailing on the massive planet. Mountains and trees would be much narrower and taller. The atmosphere would be thinner. The evolution of life on such a world might be much more rapid; provided that the rarer atmosphere did not have any unforeseeable effects, the human race would have reached much greater heights of technology than we have. Many political and psycho-

---

* We can calculate a planet's surface gravity if we know its mass and radius. It will equal the gravity at the Earth's surface multiplied by the formula $\frac{GM}{10^3 R^2}$, where G is Newton's gravitational constant of $6\cdot67 \times 10^{-8}$, M is mass in grammes, and R is radius in centimetres.

logical problems would long ago have been solved. The human body might even have evolved into pure brain. The searchers would be well-advised to be cautious before landing on a middle-aged oxygen-water planet with a mass significantly less than the Earth's. The inhabitants might be so technically advanced that we could even imagine them having the power to make themselves and their entire technology invisible at the flick of a switch.* It would be unwise to attempt to occupy a planet of this kind until the power of the inhabitants to resist invasion has been carefully gauged.

What would have happened if the Moon, instead of being located a quarter of a million miles distant, was much nearer to the Earth, let us say only 95,000 miles away? Tidal forces on Earth would be colossally increased. All our coastal regions would be uninhabitable, being flooded daily by hundreds of feet of water. One can imagine the difficulties that would face the shipping industry in these conditions. The habitable land masses would be much smaller. New York City would probably have been built somewhere in the Adirondacks, and Paris on some plateau in the Ardennes. The Earth's rotation relative to the Moon would cease altogether; one side of the world would always face the Moon, and people on the other side would never see it at all.

One could play this game indefinitely. Leaving the Moon at its present size, move it still closer, to 10,000 miles or less. Formed at this proximity, the Moon would break up into small fragments like Saturn's rings,† and at certain latitudes would on occasion obscure the Sun's light for periods of several days. Move the Earth 10 per cent closer to the Sun; we should find it covered with carbon dioxide like Venus, and with surface temperatures of hundreds of degrees. Now move it 30 per cent further from the Sun, and temperatures would be intolerably cold.

Suppose instead that the properties of some of the other Solar planets were different. What if the mass of Jupiter were increased by a factor of 1,000? With this mass the temperature of Jupiter's

---

*This is no idle joke. We should recall Arthur C. Clarke's Third Law, that 'a sufficiently advanced technology is indistinguishable from magic.'

† No large satellite can exist within about 2·45 times the radius of its primary without breaking up. This distance is known as Roche's limit.

hydrogen would be raised by gravitational pressure to 100 million ° F., and nuclear reactions would begin. The Earth would find itself orbiting between two Suns of identical size. Despite Jupiter's great distance from us, there would be an average net increase of 6 per cent in the amount of heat reaching the Earth's surface. The ice masses at the Earth's poles would probably not exist—or at least be much reduced, and there would be much more water in the oceans. The land masses would again be much smaller.

The cultural effects on humanity of this second Sun would be more serious. With two Suns in the sky, for periods of several months each year there would be no night. Astronomy would be much inhibited, cosmological knowledge and the development of nuclear energy might be delayed for ages. This idea of a world orbiting between two suns is no fantastic hypothesis; thousands of double star-systems are visible from Earth. Some fifty years ago, when the intricacies of orbital mechanics were less well understood, it was assumed that no planet—least of all a habitable planet—could maintain a stable surface temperature in a double star-system. Today, it is considered a quite reasonable possibility. The American engineer Stephen Dole, in his fascinating book *Habitable Planets for Man*, worked out the distances by which the two stars would have to be separated for a planet to orbit between them or around them without being alternately frozen and roasted. This question has been discussed on a more academic level by astronomers and mathematicians since Newton, and is usually called the 'three-body problem'. Until recently, the question posed was comparatively simple: in what different ways could three celestial bodies remain in stable orbits around each other? But the new requirement that one of them shall be a habitable planet and that the other two should be sun-type stars of roughly the same age and mass makes the question much more complicated.

Dole's answer is that the planet can be habitable in two ways. First, if the two suns are so close together that a single habitable region exists around them in which the planet can orbit. It would first take its radiation from one, and then from the other when the first was eclipsed. It is to keep this radiation fairly constant that the suns would need to be of similar age and mass. Planets in this system could also be habitable if, instead, the suns were as far apart as the Sun is from Jupiter—some 480 million miles. Habitable planets could then orbit either star without receiving excessive radiation from the other. While being unlikely to produce an advanced civilization very quickly, since the presence of two suns in the sky would inhibit the growth of science, this would be of less importance to the colonists who arrive possessing a high level of science anyway. In multiple-star systems, in which three or more stars circle or describe intricate ellipses around one another, we cannot yet assess the chances of finding planets. As David Bergamini remarks in a fine turn of phrase, 'the tempo patterns of three-star tangos, four-star fandangos and many-star mazurkas are often too complicated even for modern mathematics.'[5]

But planets in two-star systems that conform to Dole's parameters will often be suitable for colonization. Beyond the habitability of a planet, the other most important question is its possible occupation by intelligent beings powerful enough to annihilate the settlers. I have already half-humorously suggested that a super-advanced civilization might be able to make all traces of itself invisible at will, thus setting a trap for colonists. But plainly no planet will ever be settled at all if congenial appearances are always assumed to be masking a dangerous ambush. It will only be possible to weigh up the probabilities— and then take a risk. And two-sun planets, together with the very massive planets described earlier, must be considered good risks since their environments, for different reasons, would inhibit the growth of native technology.

Much mathematical speculation has been written about the number of Galactic planets, habitable planets, life-bearing planets, worlds bearing intelligent life, and advanced planetary civilizations. Readers who are not averse to arbitrary statistics, based on the flimsiest of data, may be interested in these sums. It is generally reckoned that most main-sequence stars have planets. As evidence for this hypothesis, Peter van de Kamp showed in 1969 from the minute perturbations of Barnard's Star, six light-years from Earth, that it must have a planet somewhat larger than Jupiter.[6] If we assume that this state of affairs is typical of the whole Galaxy, we can postulate that with about 25 billion main-sequence stars older than the essential 3 billion years, the Galaxy contains about 50 billion planets of all types. Suppose that 99 per cent of these are uninhabitable for various reasons, and we are left with the probability of about 500 million habitable planets.

To summarize, a planet ought in general to have the following characteristics before men land on it with a view to settlement:

1. It must have abundant plant life and wide oceans to sustain an oxygen atmosphere.

2. Its mass must be greater than 0·4 times the Earth's mass to retain this atmosphere. At the same time, it must have less than 2·35 Earth masses; otherwise the surface gravity would be intolerable. (A man would be one and a half times his normal weight on a planet of this mass.)

3. The age of its sun must be more than 3 billion years, which is more likely to be true if it has less than 1·4 Solar masses.

4. If the planet orbits in a two-star system, the two stars must either be fairly close together or quite far apart, so that the planet's orbit is stable, and the radiation reaching its surface is not too irregular.

Conditions on the planet's surface must be fairly strictly defined. Nobody would wish to live in places where wind speeds consistently exceeded gale force, blowing at more than 60 m.p.h., unless he had some extraordinary reason for doing so. A very high rate of occurrence of earthquakes, the continual falling of large meteorites from some nearby asteroid belt, or an extreme axial tilt causing very great seasonal temperature variations, could also make a planet uninhabitable. Most important of all, that minority of planets likely to be already occupied by intelligent beings should normally be avoided. Astronomers and biologists estimate the number of intelligent planetary civilizations in the Galaxy at between 1 million and 10 million, occupying between 0·2 and 2 per cent of the probable number of habitable planets. Experience on Earth suggests that men, in the long run, would have no difficulty in dealing with non-intelligent forms of life, even if some initial expeditions were wiped out by aggressive and deadly strains of bacteria.

It all sounds very easy. Imagine the orders being given in a ship in orbit round a promising star. Astronaut-professor, please aim your binoculars at that planet; tell us its mass, its surface temperatures, its wind speeds, its axial tilt, its abundance of plant life, and so forth, so that we can know whether it's safe to build a chain of cities in the temperate zone. And make sure there are no intelligent inhabitants.

At what distance from Earth today would an alien spaceship be able to detect either natural or man-made features? Cities as brilliant at night as New York and London would probably be visible through telescopes from about 5 million miles, 20 times further away than the Moon. Forests could be seen from about 300 million miles. Oceans, shining more brilliant in the Sun's light, could perhaps be seen at 700 million miles Beyond this distance, the Earth would not even be seen as a disc, but rather as a dimensionless dot, even through the largest telescope. But man's broadcasting activities in the very high-frequency range could be detected at far beyond visible limits. Some of the most powerful of all sources are the ballistic missile early-warning system radar stations, broadcasting at about 400 megahertz. With a peak power level from one of these transmitters at 1 megawatt, its signals would be detectable at a distance of 10 billion miles, more than twice as far away as the planet Pluto. The astronomer Frank Drake has calculated that some traces of our ultra-high-frequency television broadcasts could be picked up at a distance of several *light-years*, several years, of course, after they were transmitted.[7] It may be doubted whether any prospecting aliens whose computers had translated some of these broadcasts would want to come any closer.

Nothing in this or the preceding chapter should be taken to imply that manned interstellar voyages will begin before the year 2000. There are too many technical problems to be overcome, and the resources of the most advanced nations are insufficient to make the huge efforts in research and development. But by the year 2100 it will be a very different matter. If the growths of technology, science and wealth sustain anything like the rate of the last century and a half, the first tentative moves to colonize our local sector of the Galaxy will by then have begun.

There will no doubt be a dark side to interstellar travel. If journeys through Superspace prove feasible and become common-place, pollution and wars will no doubt spread to the planets of other solar systems. The bloody histories of colonial conquest on Earth will probably be repeated. Alien communities, whose main offence will be military backwardness, will perhaps be destroyed so that adventurers from Earth can fill the holds of their ships with some precious commodity. But such crimes will be rare, since any central government of these settlements will have the power to control them. A ship-full of adventurers can be pursued with equal speed by a ship-full of policemen. I assume, of course, that there will be a planet exercising central control. It is doubtful if this will always be so, since periods of anarchy seem integral to all human histories. But since all communities appear to possess either a latent or an active urge to conquer each other, it is probable that the strongest of them would reassert that central control which anarchy had temporarily abolished. Some people foresee a more peaceful and commercial life for man among the stars. 'By A.D. 7000,' says James Strong, a distinguished writer on astronautics, 'wine from the slopes of Earth's vineyards may change hands in token for curios from the hot stars of the Trapezium; rare earths exchanged for jewels or drugs from glittering Polaris, or silks and furs from Arcturus for insect-pets from far-off Wezen. There will be an upsurge of merchant adventuring such as the Earth has never before witnessed if interstellar travel gets a hold on men's minds.'[8]

The peaceful establishment of commercial markets or the deliberate destruction of inconvenient cultures by war; these are but the good and evil sides of the same concept, that of economic expansion. Both of these activities will no doubt occur during man's movement into the Galaxy, as he meets various inferior species, and we can only hope that the instances of the former will outnumber those of the latter. It is difficult to say what would happen if we met a civilization greatly superior to our own. Possibly their intellects might be so advanced that they might have evolved into pure mind, so that such ideas as economic power and the capture of stellar systems were of no interest to them. Perhaps a more interesting situation would be an encounter with a civilization of roughly the same attainments as our own, and engaged in the same expansionist policies. Suppose that each race was determined to possess one particular stellar system because one of its planets contained some vital rare commodity, or because it was a crossroads in an important trade route. Such a confrontation would be a truly major crisis for one race or the other. The loser might have to abandon all his great pretensions, and decline into insignificance. Who would prevail in such a struggle? I will end this chapter, as I began it, with Alexander the Great who, on his deathbed, gave the only possible answer to such a question. According to his biographers, the dying Alexander was asked by one of his lieutenants, 'Majesty, which of us shall succeed to your empire?' The conqueror opened his eyes for the last time and rasped out the fateful words: '*Whoever is the strongest!*' [9]

11

The Vision of John von Neumann

The schemes I have so far described, although modest in scale when measured against the colossal projects considered later in this book, will demand a level of automation far more advanced than today's. To people who know the limitations of modern computers, such instructions to a machine as, 'Land on the Moon, excavate a cave of such-and-such dimensions, seal its entrances and construct an airlock,' or, 'Navigate a voyage through Superspace to a certain star 33·5 light-years from the Sun,' will sound unreasonable. Present-day computers although much more powerful and more compact than the first-generation machines of the 'fifties, cannot perform such complicated tasks. The most sophisticated machine so far planned is a spaceship computer called ARMMS, intials for the unwieldy title, 'Automatically Reconfigurable Modular Multiprocessor System.' This machine, as its name might suggest after long perusal, is designed to operate and maintain the millions of moving parts of a manned or an unmanned spaceship.[1] It is being specially designed in America for a tour round the giant planets Jupiter and Saturn in the late 'seventies. It must operate for five years without human assistance. The task of ARMMS would be equivalent on Earth to running a major factory with neither workers nor management, doing all the production, making all the marketing decisions, manipulating the prices and writing advertisements.*

Man-made machines are becoming very much more intelligent, although, as we shall see, the limitation of size will probably make it uneconomic to try to build machines that have the full mental

* Even ARMMS is less sophisticated than STAR ('Self-testing and Repairing Computer') which would have been built for America's projected 'Grand Tour' of all the Solar System's outer planets. STAR would have functioned without human assistance for no less than 11 years. Sadly, all work on it was halted when the Grand Tour was cancelled for budgetary reasons in 1972.

powers of a man. It is now even possible to talk to machines in ordinary English, without enciphering our sentences into ones and zeros. The following conversation by teleprinter between a human and a machine, arranged by scientists at the Massachusetts Institute of Technology, demonstrates the level of intelligence reached by today's most modern machines—and their limitation.[2] The two participants, human and machine, were in different rooms since the purpose of the test was to learn whether the human *knew* if he was talking to a machine or to a fellow-man:

MACHINE. How do you do. I am the doctor. Please sit down at the typewriter and tell me your problem.
HUMAN. I have lost my luggage.
MACHINE. You say you have lost your luggage?
HUMAN. I feel lost without it.
MACHINE. Do you often feel lost without it?
HUMAN. No.
MACHINE. Why not?
HUMAN. Because I normally do not lose things.
MACHINE. What other reasons might there be?
HUMAN. My mother trained me not to lose things.
MACHINE. Tell me more about your family.
HUMAN. I have lost my luggage.
MACHINE. Did your parents ever lose you?

Sixty per cent of the people subjected to this experiment believed they were talking to a real, if rather stupid, psychiatrist in the next room. Indeed, one cannot help feeling that the machine *is* stupid, since it never attempts to consider the all-important problem of how the luggage is to be recovered. Plainly, men are going to build machines very much more capable than this one, even if there is a practical and economic limit, as I will explain later, beyond which artificial intelligence is unlikely to be increased.

That limit is still far ahead. Even household machines are becoming increasingly sophisticated. Kitchen ovens have for several years been available which turn themselves on and off at pre-arranged times. We can be awakened by a clock device which turns on the radio at any desired hour. Tape recorders built into the telephone system give us information on a widening range of subjects when we dial the appropriate numbers. The computer industry, which was purely experimental in 1945, has now grown so great that according to the predictions of a Commons Select Committee in 1971 it will be the world's third largest single industry in the 1980s, second only to cars and oil.[3] But people are too lazy even to be satisfied with a push-button world. A man presented with this Utopia will ask, 'Why can't I have a machine to push the buttons for me?' And so we have invented 'command-machines' which, using very small amounts of energy, have the function of controlling much more powerful devices. Well-known examples are the thermostat, the flyball governor on a steam engine, and the photo-electric eye which opens airport doors to approaching passengers. But people, faced with such luxuries, will ask for more, 'Why must I go to the trouble of making a thermostat? Why not get a machine to make it for me?' The final step towards real automation will eventually be made; it is nothing less than the construction of self-reproducing machines; machines which *breed* machines.

The idea of a super-sophisticated machine sounds frightening. Long remembered will be HAL, the talking computer in Stanley Kubrick's film of Arthur C. Clarke's *2001: A Space Odyssey*, which suffered a nervous breakdown because of contradictory orders from its careless human programmers. Instead of protesting to its employer that it was being asked to perform two irreconcilable actions, as a human would have done, it obediently carried out both actions—and then set about murdering all the humans in sight in order to conceal its irrational behaviour from the distant programmers. If one machine can avenge a careless error with such perverted and ferocious logic, it is not difficult to imagine the whole world covered with proliferating, murderous machines, engaged through misunderstood instructions on the task of making man extinct. A human seldom needs to be told not to eat the daisies, but a machine, by contrast, sees nothing inherently illogical in committing a lunatic action. Man will therefore find it necessary to protect himself by equipping his machines with safeguard mechanisms. If HAL had been ideally designed, he would have had built-in orders to shut down any group of systems as soon as they began to interact dangerously.

He would, at the same time, flash a signal to his programmers saying, 'Repair me! I am malfunctioning.' Such precautions as these will be essential in the early stages to the scheme for self-reproducing machines worked out in the 'forties by the celebrated mathematician John von Neumann.

Von Neumann is already well known for other extraordinary contributions to the achievements of our age. Indeed, he possessed one of the most original and penetrating minds of any age. As a colleague remarked:

> He functioned magnificently. He had the invaluable faculty of being able to take the most difficult problem, separate it into its components, whereupon everything looked brilliantly simple, and all of us wondered why we had not been able to see through to the answer as clearly as it was possible for him to do.[4]

At the Institute for Advanced Study at Princeton where von Neumann was professor of mathematics from 1933 until his death in 1957, he built MANIAC (short for mathematical analyser, numerical integrator and computer), the most advanced machine of its day, which enabled the hydrogen bomb to be built and tested much sooner than would otherwise have been possible.* He invented 'games theory', a combination of psychology and mathematical logic, from which have evolved 'war games' and those techniques of diplomacy known as 'escalation'. He suggested also that it was possible for governments to treat economic planning as an exact mathematical science instead of entangling themselves in mazes of ideological doctrine. In a work that may in time be seen to rival the contributions of Keynes, he incorporated entire systems of marketing and production into mathematical equations.[5] He proposed the use of calculus to predict statistically the probability of unforeseen or 'irrational' events, such as strikes and wars. In short, he laid the basis for a system of predicting the future over huge time-scales. He foresaw that computer technology must grow at a gathering speed. He saw also, as did few others, that man-made computers would be inadequate for

* Because of its grim work, von Neumann's colleagues protested at his plan to name his computer MANIAC. He accordingly referred to it in official literature as JOHNIAC, after himself.

extremely complicated tasks. With the experience gained from building MANIAC and several other machines, he was able to astonish an audience in 1948 with a lecture entitled 'The General and Logical Theory of Automata'.[6]

An automaton, in von Neumann's sense of the word, is a machine wholly constructed by another machine. Its potential abilities are far greater than that of a man-made computer. A man will always find it immensely difficult to build a machine that rivals him in his most innovative qualities. The reason is very simple. His own brain contains about 10 billion nerve and brain cells, or neurons. But these neurons do not *do* things. They only receive information and give orders to the roughly 10,000 *trillion* other cells in the human body. And so we are talking about making mechanical cells numbering about 10 raised to the 16th power (i.e. 1 followed by 16 zeros). Manufacturers of computers just do not have the experience to deal with units in such large numbers. Allowing for the utmost miniaturization that may be achieved, a mechanical cell, or integrated circuit (the successor to the transistor), is probably always going to be thousands of times bigger than neurons and body cells. Indeed, because of this difference in size, it can easily be calculated that a machine with the full mental capabilities of a man would need to be almost as large as Westminster Abbey.

Nor is sheer size the only obstacle to the construction of supersophisticated man-made machines. We have seen how HAL's actions became 'perverted' because of an undetected programming error. Living creatures are constructed so that malfunctions are as harmless and as inconspicuous as possible. Man-made machines, on the other hand, must be designed deliberately so that their malfunctions are as disastrous and as spectacular as possible. The reason is obvious. Living creatures can operate perfectly well despite minor malfunctions. A scientist with a twisted ankle can still work out a competent theory. But a man-made machine suffering from the electronic equivalent of a twisted ankle cannot possibly be relied on to work efficiently. Any error whatsoever represents a great risk of what von Neumann calls a 'generally degenerating process'. We would have to inquire urgently: What has caused the malfunction? What other malfunctions is it, in turn, causing? If these questions are not quickly answered, there is danger that the machine's behaviour may go from bad to worse to catastrophic. Repair becomes a far more complicated task than is surgery to living creatures. As von Neumann said in his lecture, 'We have to be far more scared by the occurrence of an isolated error and by the malfunction which must be behind it. Our behaviour must clearly be that of overcaution, generated by ignorance. Because the possibility exists that the machine may contain several faults, error-diagnosing becomes an increasingly hopeless proposition.'[7] And so 'Westminster Abbey machines', with integrated circuits representing the equivalent number of neurons and body cells found in a human being, are a technological dead-end. Constructing one would be an act of financial lunacy. Every error discovered would necessitate radical dismantling, and the maintenance costs would therefore be incalculable.

But these difficulties apply only in the case of man-made machines. As we shall see, they need not arise with *machine-made* machines, which are likely to be far more efficient. It must be accepted as self-evident that any system, whether mechanical or organic, is best equipped to reproduce its own kind. It is as idle to expect a group of men to build a perfect machine as to hope that a monkey might give birth to a horse. Major evolutionary changes cannot be expected to occur in a single generation. Men and machines are simply different species at different stages of their evolution.

John von Neumann proposed the construction of machines that would have the same reproductive ability as living organisms; his 1948 lecture showed a new insight into the way in which living creatures reproduce. He lived long enough to see his ideas brilliantly confirmed by the biologists Francis Crick and James D. Watson, who discovered in 1953 that the living cell is made of a very special substance called deoxyribonucleic acid (abbreviated DNA), which carries all the genetic information necessary for the cell's duplication and directs the building of proteins.[8] The DNA does not itself build proteins. It passes instructions for protein-building to another substance with which it is mixed—ribonucleic acid, known as RNA. RNA has been described as the 'junior assistant' of DNA. While RNA gets on with the routine and somewhat dull task of building proteins, using for its raw materials those cellular structures known as *ribosomes*, DNA does the really brilliant and imaginative work of programming its *genes*, which decide, in the case of a human baby, whether it shall have dark or fair hair, whether it will grow up short or tall, or whether its temperament will be phlegmatic or excitable. The gene's actual instructions are written on a complicated compound of DNA called *polymerase*.

A computer designer, or even a businessman running an office, may find something familiar about all this, even if he knows nothing of cell structure and has never heard of DNA. For DNA and RNA represent in their respective functions the classic system —any classic system—of efficient organization and expansion. The genius of John von Neumann gave him insight into DNA and RNA long before Crick and Watson proved that they existed. His machines, when they are built and begin to evolve, will contain reproductive systems identical in essence to those of living cells. Like a factory, they must comprise three separate components. In their simplest form, they would be something like this:

Department A, which collects raw materials and processes them in obedience to written instructions.

Department B, a message system, which writes and circulates these written instructions.

Department C, the original author of the instructions, which combines the roles of managing director (who decides what products shall be manufactured), the marketing and sales divisions (who decide what shall be done with the manufactured products), and the company architect (who must design the factory before any of these operations can begin).

It is hardly necessary to repeat that this is just as much a lesson in biology as it is in industrial organization. Polymerase, for instance, fills the roles either of the magnetic tape on a computer or of the notepaper on which the boss writes his memos. In short, Crick and Watson found that the three essential components of the human or animal genetic system were identical to those of von Neumann's automata. Department C is DNA and its assistant RNA. Department A is the ribosomes, and Department B is polymerase.

It seems almost foolish now to ask whether a *machine species* consisting of such components could ever evolve into a civilization

The answer, at the risk of triteness, is that they long ago did so, and that we are that civilization. A living organism, in short, is identical in its basic functions to the automata in von Neumanns' universal theory. The fundamental design of every micro-organism larger than a virus is, as far as we know, exactly as von Neumann said it should be.* How can a machine reproduce itelf? The one way it *cannot* is by receiving the order from its human programmer, 'Reproduce yourself.' The machine can only reply in effect, 'I cannot reproduce myself, since I do not know who or what I am.' This approach to the machine would be as absurd as if a man were to give his wife a collection of bottles and flasks containing all the chemical ingredients of one human body and asked her to build a baby. Even if the wife was a brilliant bio-chemist, she would be unlikely to produce any creature more serviceable than did Baron Frankenstein. A mother reproduces with genes, not with her hands in a laboratory. Instead, the controlling human programmer will perform three simple actions when he sets out to create a dynasty of machines:

1. He gives the machine a complete description of itself.

2. He then gives the machine a second description of itself, *but this second description is of a machine which has already received the first description.*

3. Finally, he orders the machine to create another machine which corresponds precisely to the machine of the second description, and he orders the machine to copy and pass on this final order to the second machine.

Ignoring for a moment the problem of raw materials and power supply, we can see that the man will have created by these actions a self-perpetuating cycle of machines. He has made his Adam and his Eve. Keeping in reserve his essential, self-defensive fourth order ('Destroy all machines answering to the description you now have'), he can sit back and allow the machines to carry out whatever tasks he has allotted them. The man's position is now as sublime as that of the Biblical Jehovah. If the second or third generations of his machines start to 'sin', he can destroy them with

* It is debatable whether a virus is a life-form at all. It does not reproduce in either von Neumann's or Crick and Watson's sense of the word, since it borrows the ribosomes from the cells which it invades.

a 'flood', and then start all over again with appropriate improvements. It will no doubt be necessary to perform many wrecking operations of faulty generations of machines, the equivalent perhaps of blasting Sodom and Gomorrah, before really efficient gadgetry can be built. The authors of the Old Testament seem to have had an amazing insight into this aspect of future technology.

Isaac Asimov wrote several ingenious stories about humanoid robots, machines with the exact appearance of human beings.[9] For human safety, their manufacturers were compelled by law to construct each robot so that it obeyed the 'Three Laws of Robotics'. These were:

1. A robot may not injure a human being, or through inaction allow a human being to come to harm.

2. A robot must obey the orders given it by human beings, except when such orders would conflict with the First Law.

3. A robot must protect its own existence as long as such protection does not conflict with the First and Second Laws.

This ingenious system of machine programming also reminds us of the Old Testament. Asimov's three laws are extraordinarily reminiscent of the fundamental texts in many of the world's religions. 'Thou shalt have no other gods before me', is a good example. HAL had evidently not been programmed to refrain from injuring his human 'gods'. Although there will be no need to build von Neumann machines in the likeness of human beings, it will still be necessary to prevent them from going berserk or putting us out of a job. It will probably be essential to restrict their intelligence to the purely instinctive level.

Professor Freeman J. Dyson, who teaches today at the Princeton Institute for Advanced Study, is a former colleague of von Neumann's, and he has given great impetus to the resurrection of his ideas. Dyson is a superb theorist who has worked out engineering schemes on a truly cosmic scale. He believes that self-reproducing machines will be required if man is to generate the colossal amounts of energy necessary for cosmic projects. He gave a rough illustration of how they might be used to solve an economic problem—the comparatively mundane question of how the people of Earth in the late 21st century could obtain their electric power in the improbable event of hydrogen fusion being found to be unpractical.

Dyson suggested to an audience in 1970[10] that the demand for electricity may have become so great by 2070 that nuclear fission power stations line the Pacific coast of America, taking advantage of the heat-absorbing ocean, at an average density of two per mile. These power stations poison the air with their waste discharges, heat up the ocean so that no fish live within 100 miles, and are of monotonous and hideous appearance. An alternative source of electric power is at last suggested. It is proposed that little colonies of von Neumann machines should be planted in the country's central deserts. After fierce debate, the anti-pollution lobby which detests power stations is victorious over the trade union lobby which supports the continued employment of power workers. The machines are installed. They are fuelled by sunlight, and they take moisture from the desert air for their internal needs. They break up rocks to obtain aluminium, silicon and other minerals with which to construct replicas of themselves. The output of these rock-eating automata is electricity and electrical transmission lines. Proliferating rapidly, they are able in a short time to produce 100 times more electricity than America consumes in 1970. Unlike the power stations, the machines discharge no waste heat at all and they create neither smog nor radioactivity.

Dyson himself admits that the obstacles to such a plan are formidable. The machines would certainly require rubber and cadmium to build their transmission lines, which in turn would need some means of traction to carry them from the deserts to the cities. He is not particularly serious in proposing this idea (believing that the development of fusion will make it unnecessary), and he did so mainly to illustrate his belief that von Neumann machines are best fitted to perform those extremely complicated feats of engineering of which men, using only man-made computers, would be vitually incapable.

I shall pursue no further the future problems of terrestrial power supply, since this book is concerned with the exploitation of space. In the same lecture, Dyson made another, much bolder, 'thought experiment' involving von Neumann machines. The planet Mars, he pointed out, has at present little economic value since it appears to lack both warmth and water. A group of engineers decide to remedy this without ever leaving their control panels on Earth. A rocket carrying a small colony of von Neumann machines sets out from Earth. It does not head for Mars, but goes instead to Enceladus, a moon of Saturn which consists of mixed rock and water ice. It carries an elaborate programme of instructions and some microscopic plant seeds, which will grow in the feeble sunlight falling on Enceladus. The machines will build a greenhouse from local materials for its plants. The plants will, in turn, supply construction materials and fuel to the machines.

To quote Dyson:

> For some years after the landing of the rocket nothing unusual seems to be happening to Enceladus. Then, as seen from Earth, Saturn appears to grow a new ring about twice as large as the old rings. A cloud of small objects, each launched from the surface of Enceladus by a simple machine resembling a catapult, begins to spiral slowly outwards from Enceladus's orbit. Each object has a wide, thin sail with which it can navigate in space using the pressure of sunlight.
>
> After another period of years, the outer edge of the new ring extends far out to a place where the gravitational effects of Saturn and the Sun are roughly equal. The small objects come slowly to a halt there, and begin to drift again inwards towards Saturn. This time they are not moving in spirals any more but in hyperbolic orbits. They zoom past the surface of Saturn at high velocity, receive some last-minute course corrections from computers on Enceladus, and fall free towards the Sun.
>
> A few years later, the night-time sky of Mars begins to glow bright with an incessant sparkle of small meteors The infall continues day and night, only more visibly at night. Day and night the sky is warm. Soft warm breezes blow over the land, and slowly warmth penetrates into the frozen ground. A few years later, it rains on Mars for the first time. It does not take long for oceans to begin to grow . . . There is enough ice on Enceladus to keep the Martian climate warm for 10,000 years and to make the Martian deserts bloom.[11]

This second proposal involving von Neumann machines is very elaborate, and because it requires perfect and very complex behaviour by highly advanced machines over several years, during which a single serious malfunction could wreck the whole operation, it is probably fifty years further into the future than Carl Sagan's Venus plan. But we can see perhaps from these ideas that the use of van Neumann machines in some form, the remote descendants of MANIAC, may enable us to alter the environment of the Solar System, and even to exploit large parts of the Galaxy. Man's activities in the Universe must now be envisaged on the truly grand scale.

12

Flying City-States

Before my visit to Professor Freeman J. Dyson, at the appropriately named Institute for Advanced Study at Princeton, I had built up a mental picture of him which proved to be completely wrong. I had imagined that this man who writes of dismantling planets, of reorganizing whole galaxies, and of increasing human wealth by a trillion per cent would be a sort of iron-fisted thunderer, a physical giant, whose great voice would make the window panes tremble as he bawled out incompetent students. I was mistaken on every count. Dyson is a diffident man with a soft voice who shies from personal publicity. He is only a giant intellectually. He has brooded upon man's future in the Universe on such a scale and with such grandeur that some people regard him as a madman. But his fellow scientists do not take this view. Dyson is a figure of tremendous distinction in the scientific world, a Fellow of the Royal Society, a past chairman of the Federation of American Scientists, and a consultant in numerous activities of the United States Government. The Royal Netherlands Academy has awarded him its treasured Lorentz Medal. He has won the Max Planck Medal from the German Physical Society, and in Miami the Center for Theoretical Study has voted him the J. Robert Oppenheimer Memorial Prize.

His views on the feasibility of cosmic engineering schemes evolved, slowly, as he listened to his colleagues work out their plans for detecting radio signals from hypothetical intelligent civilizations on the planets of those nearby stars which most closely resemble the Sun. Some astronomers, since the late 'fifties, have held the view that the Galaxy is filled with planetary civilizations which, being unable to visit each other physically, are busily communicating with one another across interstellar distances by radio. Frank Drake, the most famous proponent of this theory, actually experimented in 1960 to see whether he could detect any artificial signals coming from the vicinities of two nearby stars which are similar to the Sun in age and mass, Tau Ceti and Epsilon Eridani. For several weeks Drake and his colleagues tuned in to these stars with the big radio telescope at Green Bank, West Virginia. They called their experiment Project Ozma, after the fabulous land inhabited by the Wizard of Oz.[1] They used a frequency of 1420 Megahertz (a wavelength of 21·1 centimetres) for the very good reason that all clouds of interstellar hydrogen emit radio energy at this frequency. They guessed that this frequency might be familiar to all civilizations in the Galaxy and that it would therefore be the regular means of cosmic communication. After several weeks an unmistakably artificial signal was received, apparently from Epsilon Eridani. The astronomers became highly excited, and someone suggested an urgent telephone call to Washington. Soon came the anti-climax, as the artificial radio source was revealed to have been an aircraft operating from a secret military installation. The whole experiment aroused lively comment. According to Otto Struve, one of its participants,

> Project Ozma has aroused more vitriolic criticisms and more laudatory comments than any other recent astronomical venture. It has divided the astronomers into two camps: those who are all for it, and those who regard it as the worst evil of our generation. There are those who pity us for the publicity we have received, and those who accuse us of having invented the project for the sake of publicity.[2]

One of the foremost scientists to reject the philosophy behind Project Ozma was Freeman Dyson. He was not concerned with any trivial argument about whether the Green Bank astronomers had been seeking publicity; rather he questioned their assumption that planetary civilizations had nothing better to do than to transmit a continuous stream of radio messages in every direction, on the remote chance that someone might pick them up. It is conceivable that our technological activity has already been observed by alien civilizations, as I suggested earlier, and that they might be trying to speak to us. But it is just as probable that all signs of our existence may have been hidden from them by the sheer scale of interstellar distance, or blotted out by the radio halo of the Sun.

If this is so, aliens who *did* suspect (a) that the Earth existed, (b) that it harboured intelligent life, (c) that such life-forms would benefit from their missionary teachings, and (d) were too naive to realize that such education can be dangerous for the educators, *might* beam radio messages to Earth. Yet they would still have no means of knowing *when* the people of Earth would be ready to receive them. After all, we have only had radio astronomy for 25 years. And if these aliens were enthusiastic missionaries, why should they confine their attention to Earth? There will be dozens of likely stars as close to them as we are. To have any chance at all of making contact, the aliens would have to transmit in thousands of different directions simultaneously; not knowing when their pupils-to-be would become technically literate, they would have to continue these transmissions for millions of years. The proposition becomes still more doubtful when we consider the amount of energy needed to send radio messages across interstellar distances. It is hard not to accept Dyson's conclusion that the aliens would think of more profitable ways to invest their wealth. Rather, he suggests, they would concentrate on expanding their own technology. Consequently, he has urged astronomers interested in alien civilizations not to expect messages from them but rather to seek visible signs of their technology. As he wrote in a memorable paper in 1959:

> It seems a reasonable expectation that Malthusian pressures will ultimately drive an intelligent species to adopt a more efficient exploitation of its available resources. One should expect that within a few thousand years of its entering the stage of industrial development, any intelligent species should be found occupying an artificial biosphere which completely surrounds its parent star.[3]

This paper represented the intellectual birth of that gigantic engineering idea known as the Dyson Sphere. Aliens who deliberately dismantled their largest planet, their equivalent to our Jupiter, would exploit the whole radiation from their parent star by placing the fragments of the broken planet in convenient orbits around it. They would then use these new worlds for residential or mining purposes as they saw fit. The 'sphere' would not be a rigid construction, since that would be dynamically impossible, but would consist rather of a loose swarm of hundreds of thousands of loose-flying objects all orbiting in the same direction, and would be similar in its mechanics to the famous Rings of the planet Saturn. Saturn is orbited by hundreds of millions of tiny boulder-like fragments of ice and dust, so that the planet appears from telescopes on Earth to be surrounded by a solid ring. The Dyson Sphere will from a distance resemble such a ring, except that it will be very much broader and thicker in proportion to its sun than the rings are to Saturn.

The 'sphere' will never be total, since the objects which comprise it must all fly in the same direction to avoid collisions. The 'top' and the 'bottom' of the central sun will therefore be clear of all objects, and would be visible from outside. Yet an outside observer, looking at this arrangement from an interstellar distance, would find the light of the sun in question to be strangely dimmed by the obstructions of the sphere; instead of the white or yellow colour which its age and mass dictate that it should be (and still would be if viewed from inside the sphere), it would emit light mainly in the infra-red. This would be stellar radiation in other words, most of whose energy would already have been tapped by the people living near it, and of which only the infra-red 'dregs' would be allowed to escape into space. This is a simplistic way of putting it, but without giving a complicated lecture on optics, we can say simply that when the emission of white or yellow light is substantially disrupted, from very far away it will appear to veer into the infra-red end of the spectrum. Dyson therefore advised the astronomers not to expect too much in their search for alien messages.[4] They should rather examine with suspicion any infra-red stellar objects in the Galaxy which seem to defy natural explanation.

Let us have done with these aliens. To observe them or make contact with them would be a bonus for our civilization which we have no right to expect. I have introduced them merely to show how Dyson and his Russian colleague N. S. Kardashev came to work out their ideas about the feasibility of a Dyson Sphere, or, as Kardashev calls it, a 'Phase 2 civilization'. Let us return to Dyson's original statement that: '. . . within a few thousand years of entering the stage of industrial development, *any intelligent species* [my italics] should be found occupying an artificial biosphere around its parent star.' This definition plainly includes us. If it is true, as I have argued, that there is no natural limit to technological growth, then it is evident that we will eventually be compelled to dismantle the giant planet Jupiter in order to construct a Dyson Sphere around our Sun. We shall do this for the obvious reason that only in the fragments of the giant planets shall we find sufficient living space for our expanded population* and raw materials for our industrial operations.

What kind of society would be capable of such feats? In knowledge, wealth and power it would differ from ours as ours differs from that of the Middle Ages. But in what precise measurable quantity do these societies differ from one another? The phrase 'level of technological activity' is not just a piece of economists' blarney. We can actually measure this activity in terms of the amount of energy which a society deploys. The basic unit for measuring energy is the erg, and we can say that a society's level of wealth and power is best measured by the approximate total number of ergs it deploys each year. I will give briefly some examples of 'ergometrics' and the fairly simple arithmetic that goes with it.

An erg is the amount of energy required to push one gram of matter for one centimetre. A tiny insect, when walking along the floor, therefore deploys 1 erg for every step it takes. A man climbing a staircase deploys 1 billion ergs on each upward step. These actions represent relatively small amounts of energy, and so when we speak of whole societies we are soon involved with very large numbers. I had hoped to get through this book with American 'billions' and 'trillions', but convenient though these words are, they are of no use in astronomical ergometrics. Let us try instead with powers of 10, the simple mathematical method which scientific journals use to describe almost any number, great

---

* This is not meant to contradict the prediction in Chapter 2 that the world population will stabilize at about 10 billion in the mid-21st century. 'Stabilization' means the end of an urgent exploding population crisis, but it does not mean that there will not still be a 'creeping increase' in the order of 0·4 to 0·5 per cent a year. Human colonies on the Moon, Venus and possibly Mars will at first probably expand their numbers very rapidly. The *human* population, as opposed to the Earth population, will, over the course of centuries, show an inexorable increase.

or small. These journals seldom talk of a 'billion'. They say instead 10 raised to the power of 9, or 10^9.* This means the figure 1 with 9 zeros added to it. Thus 10^1 is 10, 10^2 is 100, 10^3 is 1,000, 10^4 is 10,000, and so on.

It is very easy to multiply and divide with these powers of 10. To multiply two numbers, we add the index figures, and to divide we subtract them. Thus, to multiply a billion by 100,000 is to multiply 10^9 by 10^5. We add together the 9 and the 5, giving the correct answer of 10^{14}. To divide a billion by 100,000, we subtract 5 from 9, making 10^4. Let us try out the system.

The Roman Empire, in its most energetic phase, with its projects of road-building, public works, and the movement of its fleets and armies, deployed an *annual* budget of about 10^{24} ergs. Our global civilization, with all its factories and road, rail, sea, air and space vehicles, deploys an annual budget of about 10^{29} (i.e., 100,000 times more energy per year than the Roman Empire). Here are some other examples; the explosion of 1 ton of T.N.T. unleashes in a few seconds about 10^{16} ergs. The Saturn Five Moon-rockets needed 10^{22} ergs to lift them into space. A hydrogen bomb of, say, 10 megatons releases 10^{23} ergs, the amount of energy which the Romans would have used in just over a month. The great volcanic explosion at Krakatoa in 1883, which killed 36,000 people and was heard 3,000 miles away, unleashed about 10^{25} ergs, 100 times more energy than a 10-megaton hydrogen bomb. Some flares from the surface of the Sun, which cause magnetic storms on Earth, are themselves the result of outbursts of about 10^{31} ergs. And the Sun itself each year produces 2 times 10^{39} ergs.†

What *is* 'technological growth', the actual agent which raises the levels of technology? Many people do not understand its real nature; if they did, they probably would not sneer at it so much.

* To express an intermediate number, like 9,650,000, we would say 9·65 times 10 to the power of 6, or $9·65 \times 10^6$.

† This energy is nothing when compared with the real violence of the Universe. A nova explosion unleashes 10^{44} ergs, a supernova 10^{49}, a chain-reaction of supernovae of a super-supernova, 10^{54}, the explosion of an entire galaxy about 10^{58}, a 'quasar' about 10^{61} in its total lifetime. The 'big bang' which created the Universe would have been in the order of 10^{80} ergs, which, by a strange coincidence, is also the approximate number of atoms in the known Universe.

It has little directly to do with pollution, or the rate of construction activity, or the growing number of scientists and engineers, or the quantitative growth of air freight. It does not advance in a steady flow but rather in a series of very large jumps. A technological breakthrough occurs when somebody discovers a way to tap energy at a cost of less effort than before. There is plenty of energy in the Universe, but human progress has sometimes been slow because inventors have been unable to find practical ways of harnessing it. The same technological breakthrough cannot occur twice. The first wheeled vehicle, built some thousands of years before Christ, was a tremendous development which altered the whole structure of demography and commerce. But the second wheeled vehicle was, well, just another wheeled vehicle. It was probably a much better vehicle than the first—but it was the first that made the breakthrough, and it was the first which directly caused a growth of human civilization greater than the world had ever known. So it is with every useful advance.

We at present await two very important technological developments. The first is the space shuttle, which will reduce the cost of launching men and material into orbit by about 80 per cent, thus enabling us to deploy far more energy in space; the second is hydrogen fusion, which will eliminate all problems of energy scarcity and radioactive wastes. Each of these improvements will lead to very large jumps in our annual energy budget. The space shuttle is due to begin operations in 1979, while it is believed that hydrogen fusion will be supplying at least some of our commercial electricity by 2000 or 2010.*[5] Within a few decades of their becoming widely used, the combination of these two breakthroughs is likely to raise our energy budget by a factor of 100 or 1,000 to 10^{31} or even 10^{32} ergs. In the latter event, we shall be in the proud position not only of having largely mastered our pollution but of being 100 million times richer in energy than the

* It is possible that even before 1990 we shall learn to imitate the Sun in using hydrogen to produce controlled energy by fusion. The main problem has been to sustain temperatures of about 200 million degrees F. for more than a few hundredths of a second. But fusion research is going on in several countries, and it appears inevitable that fusion will ultimately be our principal source of electricity supply. The deuterium in the oceans, the raw material for fusion, can supply man's energy needs for hundreds of thousands of years.

Roman Emperors and a good deal more civilized. As Sir Fred Hoyle puts it,

> I would take to task those governments, those industrialists and economists, who see present-day technology as being more or less the end of the road; a road that started with our remote ancestors swinging about in trees, a road which has taken us precisely to our present-day position, and will take us little or no further. The truth, I believe, is exactly the opposite. The gap between us and the civilizations of the future may be as large as a factor of 10^9. The road is only just beginning. Almost everything remains to be played for.[6]

Three things, knowledge, wealth and power, form the key to human progress. Each makes the other two possible, and each, by increasing, causes the other two to increase. With the wealth of the industrial nations now increasing by more than 4 per cent a year, and tending to double during each 17-year period while this average annual rate is maintained, accomplishments which now are uneconomic will become quite normal. By the latter part of the 21st century, manned flights to Mars and Venus are likely to become as commonplace as a flight across the Atlantic today.

But why, some people will ask themselves, is it necessary to land on Mars or Venus? Why is it necessary to land anywhere at all? Is there any absolute reason why everyone should be permanently resident on the surface of a planet, as opposed to a flying city? I have described the communities that will be living on the Moon and Venus, but in each case these colonists will suffer some inevitable disadvantages from their environments. Under the constant threat on Earth of overcrowding, civil strife and nuclear war, and above all, because of the commercial need for space manufacturing in Solar-orbiting factories, the 1990s are likely to see the beginnings of the construction of flying cities, which will ultimately expand to many miles in diameter, and provide permanent homes for hundreds of thousands of people.[7] I have already mentioned the Skylab orbiting space station in which astronauts occupied quarters as large as an average family house. They lived almost as comfortably as we do on Earth. I say 'almost as comfortably' since they had to acclimatize themselves to weightlessness. Skylab was too small a station for it to spin on an axis and thus provide gravity artificially for its human occupants. But imagine instead two Skylabs of similar mass; they are attached by a rod some 300 yards long.* The astronauts in each of them desire gravity, so that they can walk about normally as they would on Earth. The solution is simple: the astronauts in each station fire small directional rockets simultaneously and in opposite directions; the whole system at once begins to spin, and brings gravity to each station. Suppose that the astronauts made a mistake and fired their rockets for too short a time, thus producing a gravity too weak for comfort; the answer would be simply to fire them again, and keep on firing them until gravity was adequate. But what if they fired their rockets for too long? Gravity might then be too strong, and each man would complain that he weighed half a ton. This problem also could be solved. The rockets would be fired in the opposite direction until gravity in the stations was at an agreeable strength. The entire system would maintain that gravity indefinitely.

I speak somewhat lightly of a 300-yard connecting 'rod'. But I only use this idea to show that weightlessness will *not* be a problem in space travel. It is easily removed by causing a space system to spin. The much larger space stations that will succeed Skylab are likely to be given this initial spin if their inhabitants are to live in comfort. The McDonnell-Douglas Aircraft Corporation, looking beyond 1990, have studied the concept of a 'third generation' commercial space station that would carry 400 people.[8] In a giant construction like this there would be no need for such rods. The entire system would be large enough on the human scale to spin on its axis and provide gravity in every part of it except the absolute centre. Nor, once this gravity had been provided, would there be any reason to restrict the size of the station's personnel to 400. The stations will be immensely expanded in size. They will become true space cities, many miles across and accommodating tens of thousands of people. Because of the gravity system, the outer perimeters of these cities will be residential, while the

---

* The system must be at least 300 yards long, otherwise the astronauts would be pulled by gravity in two opposing directions, both outwards and inwards, which would result in nausea. This would be due to gravitational 'Coriolis forces', which in a relatively small satellite like Skylab would negate the effect of spin. The Coriolis effect only becomes negligible in a much larger satellite.

sections around the centre, where weightlessness is unavoidable, will be kept for industrial purposes. Even more splendid amenities can be introduced. If the cities are shaped like flying discs, they can have a dual, or complex rotation that exposes each side of the disc to the Sun for 12-hour periods, simulating the Earth's equatorial days and nights exactly. After each period of 12 hours, the side of the disc facing the Sun automatically swings round and faces the darkness of space.

The one-sixth gravity in the surface of the Moon will give us some new industrial processes in partial weightlessness. But weak gravity is no substitute for zero gravity. While we look to the Moon for our greatest advances in vacuum technology and astronomy, it is only in the weightless environments of the centres of space factories and space cities that really sophisticated advances in weightless manufacturing will be achieved. These will include the growth of pure crystal structures for scarcely imaginable electronic applications, high-strength extremely light metals, methods of nuclear engineering that would be too dangerous to conduct on Earth or even on the populated Moon, and, paradoxically, metallurgical developments in conditions of extremely high gravity. These conditions can be created by spinning an industrial satellite extremely fast, so that objects on the inside surface of its sphere can be hurled outwards with terrific centrifugal force. All such techniques will contribute to a way of life everywhere in the inner Solar System that would seem to us like magic.

The inhabitants of flying cities could reproduce almost any Earthly conditions they desired. Some external parts of their structure would not rotate, but would for ever face the Sun. There would be great fields of agriculture or hydroponic farms beneath transparent domes. Imagine a series of these cities, shaped like gigantic wheels, each 20 or so miles in diameter, and each with huge docking bays for their commercial and inter-city transport activities, all circling the life-giving Sun in varying orbits. Each city would be orbited by many satellites smaller than itself. Among them would be the fast spaceships that could carry men and material to other cities and to distant planets. They would orbit around their parent city until they were wanted, just as on Earth a fast motor-launch is tethered to an ocean liner as she rides at anchor.

An alternative shape for these flying cities has come in a spectacular proposal from a Princeton physicist, Professor Gerard K. O'Neill. He suggests that cities shaped like giant cylinders would be even more convenient and practical than discs. His idea has attracted much favourable attention among scientists, and I will summarize it here.[9] The largest of O'Neill's cylindrical cities would be about 20 miles long and five miles wide. Tens of thousands of people would live on its curved interior. This interior would be covered as desired by lakes, hills, forests and meadows. The Sun would power the city's steam engines, thus providing pollution-free energy. A huge mirror outside the cylinder would slowly open and shut on a pivot each day, making the complete movement in 12 hours. The Sun's image and warmth would be reflected by the mirror through a great Solar window, giving the impression that it was rising, crossing the sky, and setting, just as it does on Earth. As O'Neill himself remarks:

> By 2074 a large part of the human population could be living in space colonies, with a virtually unlimited clean source of energy for everday use, an abundance of food and material goods, freedom to travel and independence from large-scale governments.[10]

Despite O'Neill's idyllic view, it may be objected that it would be intolerable to live permanently in one of these flying cities. In the early 400-man stations, life will indeed be very constricted, and few people will want to stay in them for tours of duty of much longer than several months. But as much larger cities are built there need be few limits to the size of parks and gardens that can be added to them. Every plant or tree in these parks, growing beneath gigantic transparent domes, will be as real as their counterparts on Earth. There is no absolute limit whatever to the size to which these cities can grow. In any case, cities on Earth suffer from problems which are mainly the fault of their own geography. The centre of Los Angeles lies in a mountain bowl which, during certain weather conditions, concentrates the smog from car exhausts to a barely acceptable level. Manhattan, the main island of New York, is so constricted that a large number of its citizens live in de-humanizing skyscraper apartments, and the growth of slums is endemic. The mountain ranges of Japan have forced its industry to concentrate in a huge area around Tokyo. The amount of resulting atmospheric pollution in Tokyo can be judged by the fact that in recent years policemen have had to wear smog masks while on traffic duty. Washington, D.C., capital of the Western world, suffers from such fearful summer humidity that until the invention of air-conditioning British governments regarded their embassy there as a hardship post. It is not so much the behaviour of the modern inhabitants of these cities that should be blamed for these problems, as some writers have argued; it is rather their geography. Cities on Earth are afflicted with countless difficulties, which their founding fathers, being ignorant of the effects of modern industry, could never have foreseen. It is no answer to build more big cities on Earth, which would merely rob other people of their open spaces. The only ideal answer will be to build new and limitless cities in the almost limitless vastness of interplanetary space. They will be able to offer every amenity which a city should have. Space travel to and from them can be as simple and as cheap as air travel between Earthly cities today. Their industrial wastes can be ejected into space in special containers, and propelled so that they are devoured by the Sun, thereby inconveniencing nobody. Their basic electric power source will also be the Sun, and there will be no clouds to interfere with it.[11] Every disaster in these cities will be the fault of its inhabitants, and if anything breaks down there will be no one to blame but the Mayor.

As the cities become more complex there is no reason why they should always themselves be satellites of the Earth. Indeed, the people of Earth would have a powerful argument that they should not be so, since a sufficiently large number of them would block off the light of the Sun. They are likely to follow their own orbits round the Solar System, some of them vanishing behind the Sun for many months. Krafft A. Ehricke sums up: 'With their giant factories and food-producing facilities, the cities will maintain their own merchant fleet of spacecraft, their own raw material mining centres on other celestial bodies, and be politically independent city-states.'[12]

Construction of the cities will begin, probably in the last decade of the 20th century, with extremely lightweight metal girders made possible by Earth-orbital technology. The development of

Lunar industry in the following decades will make it still easier and cheaper to obtain lightweight construction materials. But as the 21st and 22nd centuries proceed, the cities will become ever more numerous and grandiose. Gradually, there will come a crisis of raw materials, which factories on the Moon will no longer be able to supply in sufficient quantity. If the cities are to continue to grow, they will urgently need to exploit some of the loose matter in the Solar System, which fortunately will be ready to hand. Between the orbits of Mars and Jupiter, making up a broad swathe about 200 million miles from the Sun, is a huge swarm of tiny planets known collectively as the Asteroid Belt. All are far too small to hold an atmosphere. The largest of them, Ceres, is no more than 430 miles in diameter. The rest, numbering about 50,000, range in size from 300 miles across to mere lumps of rock a few feet wide. Added together, they have a mass of about 1 per cent of the Earth's. Since they are of no economic value in their present position, and because their scientific interest will have long been exhausted by the mid-22nd century, many of them will be broken up to provide the material for new urban construction. In some cases, they will be broken into pieces by explosives, so that the fragments can be propelled by rocket engines into more convenient orbits nearer the Sun. Some of the larger ones may be hollowed out so that they themselves *become* cities. Smaller ones can be similarly hollowed out and used as giant cargo vehicles. As the Russian rocket pioneer Konstantin Tsiolkovsky once predicted, 'We shall one day learn to ride the asteroids as today we ride horses.'[13]

By the 23rd century, a region approximating to the Earth's orbit round the Sun will be a seemingly endless stream of massive residential and industrial hardware. Seemingly endless? To a visitor from our time, this would indeed appear the case. But the people of that age will not be satisfied. They will be a civilization with an energy budget of about 10^{37} ergs, wielding a power 100 million times greater than ours today. But, as always, they will be worrying about their future. They will be talking about a crisis of raw materials. They will look with indignation at huge gaps in the vast ring of factories and cities round the Sun. 'Look at all that black emptiness!' we can imagine them saying. 'New Chicago went behind the Sun a week ago. New London is not due to appear for another four days. What is there between them? A few insignificant villages and amusement parks, the largest of them scarcely 10 miles in diameter. All that empty space is being wasted. It ought to be *doing something* for mankind. But what can we do? We want to expand but we cannot touch the interior planets since they are covered with people. People are crying out for living space, and industrialists and governments are screaming for raw materials. Where shall we obtain them?'

There will be one place to obtain them, fantastic though the suggestion may seem, even to the people of the 23rd century. The giant planet Jupiter is only twice as far from the Sun as from the asteroids; what economic value does it have in its present orbit? our descendants will ask. It is such a tantalizing lump of mass, 30,000 times more massive than all the now-vanished asteroids. If it were taken apart, mankind would have no more worries for thousands of years. The people of the 23rd century, the people armed with 10^{37} ergs, will begin to examine Jupiter, the giant of our Solar System, with an acquisitive and predatory eye.

13

Building the Giant Sphere

We have already learned a great deal about the Universe from watching Jupiter and its system. Galileo, in 1610, observed its four largest moons in orbit around it—the first observational evidence that small celestial bodies orbit larger ones, and that the Sun must be the true centre of the Solar System. In 1671 the Dane Olaus Roemer used the different times of the year when Jupiter's moons were seen to be eclipsed—depending on whether the Earth was approaching or receding from Jupiter—to make the first roughly correct estimate of the speed of light.* A German astrologer now tells us that those born under Jupiter are inclined to 'expansion and enlargement'.[1] His view has a certain irony, since the logical climax to a policy of expansion and enlargement is to destroy the very source of its influence and to hurl this monster of the Solar System from its orbit. Jupiter really is a monster planet.[2] Its mass is more than twice that of all the other eight planets put together. It occupies 1,300 times the volume of the Earth; it has 318 times its mass and 11 times its diameter. It drags along a dozen moons on its 12-year orbit of the Sun, of which two are larger than our own. Everything about Jupiter is on the titanic scale. Our entire Moon could be hidden inside its atmosphere.

There is not the slightest possibility that humans, or any form of life we know of, could live on its surface. Its atmosphere consists of unbreathable hydrogen, ammonia and methane. So strong is its gravity that a man who weighed 150 pounds on Earth would weigh nearly a quarter of a ton on the surface of Jupiter. But he would never reach that surface. Any spaceship that penetrated the atmosphere, no matter how sturdily it was built, would be

* Roemer's estimate of the speed of light was in the right order of magnitude, 141,000 miles per second, compared with the 186,000 m.p.s. figure generally accepted today .

crumpled like tinfoil by the pressure of gases. The surface itself is in all likelihood lashed by torrents of ammonia rain. Judging by radio bursts which reach us from Jupiter, these downpours are triggered by thunderstorms of unimaginable violence. A single one of these bursts, lasting one second, releases radio waves equivalent in energy to those generated by 100 billion Earthly strokes of lightning. During a storm, these bursts follow one another in staccato succession, giving us some mental picture of a world so cataclysmic at the surface that superlatives lose all meaning.

The very 'surface' of Jupiter is mysterious. Astronomers have reached only a tentative agreement about its nature. Some 2,000 miles inside the atmosphere, there is a layer of solid hydrogen, or hydrogen 'slush'. This would be liquid hydrogen, mixed richly with rocky materials which solidify more readily than itself. Any large object which found itself resting on this mushy 'surface' would not remain there for long, but would sink rapidly, as through a quicksand. It would stop sinking after some 5,000 miles—perhaps more or less; the internal geography of Jupiter is extremely vague—when it reached a new surface, this time of almost impenetrable solidity. Suddenly, at some point between 7,000 and 20,000 miles below Jupiter's visible gaseous layer, there is an inner core of dense metal. * This is metallic hydrogen, an extraordinary substance unknown to us on Earth. No one has ever seen it, yet under sufficient gravitational pressures it must exist. When the pressure of a planetary atmosphere rises above 1 million Earth atmospheres (1 atmosphere is 14·7 pounds per square inch), hydrogen atoms are subjected to such enormous compression that their electrons are separated from their nuclei. In this state, hydrogen becomes a metal.

Here, then, is a world with a spherical core of metallic hydrogen some 75,000 miles in diameter, covered with a layer of hydrogen 'slush' some 7,000 miles deep, which in turn is covered by about 2,000 miles of thick gas. Yet this is not all. According to the spectroscope, that marvellous telescopic attachment that breaks down the light from celestial objects into different colours, thus enabling us to analyse their composition, only 78 per cent of

* These are no great distances in comparison with Jupiter's diameter of 89,000 miles.

Jupiter consists of hydrogen. Helium comprises 10 per cent, and the remaining 12 per cent is oxygen, nitrogen, carbon, silicon and aluminium, together with iron and other heavy elements, and taking the forms of rocks and ice. These are the principal elements which, in their various mineral formations, make up the Earth's crust and mantles. In Jupiter, therefore, we have sufficient mass in the required elemental forms to build about 38 new worlds with the mass of the Earth, or 3,000 worlds with the mass of the Moon, or about 300,000 flying cities each with the mass of Ceres, the largest of the asteroids—or any desired combination of these.

Some people have objected with horror to any plan to dismantle Jupiter. The giant planet, they assert, has a measurable influence over the orbits of all other planets. Disrupt that influence, they say, and the entire harmonious mechanics of the Solar System will be ruined. They do not know what precise effect Jupiter's disappearance will have on the Earth's orbit round the Sun but, they have informed me with many a wagging finger, there is bound to be an effect, and it will destroy us. What effect? I ask. To this question they confess an ignorance of mathematics, but insist, somewhat mysteriously, that you cannot interfere with nature on such a scale without paying a terrible price. Either the Earth's mean orbital distance from the Sun will be fractionally increased, which could trigger a new Ice Age and expose the temperate zones to the onslaught of glaciers, or else decreased, thus producing the opposite, but equally deadly, effect in which the polar caps would be melted by the greater Solar heat, and flooding all coastal cities. Critics with a genuine astronomical background sometimes conclude their lectures to me by saying that the Solar System 'consists of Jupiter plus debris'. Remove Jupiter, and who can foretell the fate of the debris?

The answer is that Jupiter's removal will have no appreciable effect on the Earth's orbit at all. As the mathematician and astronomer Iain Nicolson makes clear in an appendix to this book, it seems quite certain that no drastic consequences would ensue.* Jupiter is simply too small and too far away from the

---

* The appendix by Iain Nicolson, lecturer in astronomy at the Hatfield Polytechnic Observatory, which he wrote specially for this book, appears on page 197. It is a technically-written but nonetheless fascinating inquiry into the most efficient means of constructing a Dyson Sphere.

Earth for its disappearance to make any but the most insignificant difference to us. Five times nearer to us than Jupiter is there is a body whose mass is a thousand times greater—the Sun. The Solar System ought more properly to be described as 'the Sun plus debris'. The Sun has 333,000 times more mass than the Earth, and it contains 99·86 per cent of all the mass in the Solar System. As Nicolson points out, it exerts 16,000 times more gravitational force on the Earth than does Jupiter. These figures appear decisive, and dispose altogether of any suggestion that Jupiter is essential to the inner Solar System's equilibrium.

Yet the big planet has at present a very great scientific value, and many questions about its composition need to be answered. Even if it were economically possible to dismantle it today, today's astronomers would fervently oppose such a scheme. Although the Jovian atmosphere would crush to pieces the scientific instruments of all spaceships that entered it, much will be learned from all the manned and unmanned spacecraft that will fly around Jupiter within the next century. It is likely that a manned scientific observatory will eventually be placed on Amalthea, Jupiter's innermost moon. At this close distance of 110,000 miles the giant planet would fill a quarter of the sky. All kinds of new scientific instruments, with a sophistication far greater than ours, will probe the murky cloud-banks of Jupiter from the desolate, rocky vantage-point of Amaltheia.[3] But even these pioneers on their lonely outpost will be unable to *see* the slushy surface of Jupiter, let alone the metallic surface beyond. Long before the 22nd century is over, therefore, Jupiter intact will have yielded all the scientific fruits that it can. By the 23rd century, even the astronomers will be interested in seeing it from the inside, and they can be expected gradually to drop their objections to dismantling it. If not, humanity's ever-growing demands for new raw materials and territory will overrule them.

What will prove the most efficient method of dismantling Jupiter? By 'efficient' I mean the method which achieves its purpose in the shortest possible time, with the minimum expenditure of energy and with the least danger to man, distributed as he will be on the Earth, the Moon, Venus, and probably Mars as well as the innumerable flying cities. 'It is possible to take planets apart,' wrote Dyson in 1966.[4] To prove this, he published a paper showing how Jupiter could be taken apart even in terms of today's technological knowledge. The task, by this system, would take 40,000 years to complete, and would involve girding selected latitudes of Jupiter with metallic grids until the surface of the planet was wired like the armature of an electric motor. Solar energy, concentrated by enormous reflecting satellites, would create sufficient electrical stresses through these grids to increase the planet's speed of rotation until centrifugal forces began to tear it apart. Jupiter rotates on its axis about once every 10 hours. If this rotation were accelerated to a rate of once every hour, stresses would build up until the whole region around the Jovian equator would rip itself loose and take off into space at escape velocity. As the rotation speed increased still further, more parts would fly off. What once had been a planet would be a large collection of moons orbiting round a body only slightly larger than themselves.

Dyson was not implying that this would prove the best method. He was anxious only to show that the dismantling of Jupiter was not a fundamental impossibility and that it could be accomplished. His grid-encircling method has two major disadvantages. It would take 40,000 years (by Dyson's calculation) before a single fragment of useful material could be obtained. It is hard to imagine any society, however wealthy, investing significant sums in an enterprise that would yield no profit until 400 centuries had elapsed, and in any case only 38 Earth masses at most would be obtained. Of the rest of Jupiter's mass the 78 per cent consisting of hydrogen would be wastefully dissipated in space with the removal of that mighty gravitational force which had compressed it into a metal.

A more efficient method is needed which would convert even hydrogen into useful material. Perplexed by this problem, I consulted Iain Nicolson, who concluded that the only means of doing this is through the process of thermonuclear fusion, which takes place naturally in the interior of stars. Our discussions, by determining today the best means of constructing a Dyson Sphere, were aimed at discovering how it *will* be constructed when the time comes. It is likely that within four or five centuries we shall be capable of devising explosives that would shatter Jupiter in a matter of hours. Such a bomb, incidentally, would need the energy equivalent to 10^{21} hydrogen bombs each of 40 megatons. The difficulty would be that after such a huge explosion the fragments of Jupiter would scatter throughout the Solar System, and many would receive such boosts of acceleration that they would head out irretrievably into interstellar space. Others would fly towards the Sun. They would intercept the Earth's orbit and perhaps crash into the Earth. The explosion method, therefore, would be unprofitable and dangerous.

The Birmingham physicist J. H. Fremlin first suggested to me that fusion reactors (unmanned, obviously) should be placed in sub-orbits inside Jupiter's atmosphere, flying above the slushy surface.[5] They would suck in the gaseous hydrogen and convert it steadily by fusion into heavy elements such as iron. Rocket motors, or more probably magnetic fields activated by synchronous satellites, would extract the iron from Jupiter and propel it across interplanetary space until it reached a proximity to the Sun equivalent to the Earth's orbit. The huge iron fragments, some of them as large as the Earth itself, would then orbit the Sun in the same direction as the Earth, becoming ever more numerous as the reactors did their work and the Dyson Sphere took shape. Nicolson endorsed Fremlin's plan but pointed out one serious danger. Nuclear fusion not only converts hydrogen into heavy elements, but also unleashes huge amounts of energy. Parts of Jupiter would become heated into a nuclear furnace. As Nicolson puts it, 'if all the radiation produced in these reactions were radiated isotropically (i.e. equally in all directions), the Earth would intercept from the direction of Jupiter a quantity of radiation equivalent to 4 per cent of the radiation now received from the Sun, possibly in the form of dangerous short-wave radiation'.

There would be little point in building a Dyson Sphere if half the human population received leukemia from the industrial wastes of its construction. Nicolson has solved this difficulty with the suggestion that a huge screen, a sort of mini-Dyson Sphere, should be erected around Jupiter itself while the reactors are working. The screen would intercept Solar energy to provide the power supplies for the reactors. Where would we obtain the materials to build this screen, since little can be extracted from Jupiter without unleashing dangerous short-wave radiation? The screen does not have to be particularly massive. In fact, it can fairly easily be built from a combination of the largest moons of Jupiter, Saturn and Neptune, three of which are larger than our own Moon, and *one* of the inner planets. Which inner planet? The question will be asked with some anxiety in that age, since most of them will be carrying large human colonies. Happily there is one planet which may never be permanently occupied, since its closeness to the Sun makes its surface intolerably hot. This is Mercury, which is only 36 million miles from the Sun, and more than four times more massive than our Moon. A sequence of big hydrogen-bomb explosions at well-chosen points on Mercury's surface could break the planet loose from its orbit, and send it spiralling out towards Jupiter. Once having arrived in Jovian orbit, this rocky, almost airless world could then be fairly quickly dismantled, either by Dyson's grid method or by surface mining. With an atmosphere of carbon dioxide 1,000 times thinner than the Earth's and a low escape velocity (3 miles per second against the Earth's 7), there should be no difficulty in dismantling it rapidly. When the broken-up mass of Mercury is added to those of Jupiter's giant moons Io, Europa, Ganymede and Callisto, together with Saturn's Titan and Neptune's Triton, the result will be a total mass of 16 per cent that of Earth's. This will be sufficient to place around Jupiter a mini-Dyson Sphere, protecting the Earth and the inner planets from lethal bursts of nuclear radiation. The protective sphere, again resembling a thicker and broader version of Saturn's rings, will orbit Jupiter on the Solar ecliptic, so that it screens all parts of Jupiter from the inner planets. This will be more easily achieved if the protective Sphere orbits Jupiter at a synchronous distance, so that each part always rides over the same point on Jupiter's surface. A synchronous orbit means a distance from Jupiter of some 37,000 miles,* and so we are talking about hundreds of thousands of loose-flying objects forming a semi-sphere which is nearly 160,000 miles in diameter and 500,000 miles in circumference. Meanwhile, as materials for a full-sized Sphere are extracted, they will be ejected through the gaps in the mini-Sphere and placed in suitable orbits round the Sun.

* Because of a Newtonian formula based on a planet's period of rotation, which in Jupiter's case is just under 10 hours.

Within a few centuries after this great work has begun, a much greater Sphere will circle the Sun. It will comprise tens of millions of loose-flying objects, ranging from several hundred Earth-sized worlds to countless smaller industrial planetoids. But it will always be desirable to build Earth-sized planets with at least twice the mass of Mars, for these alone will be sufficiently massive to retain breathable external atmospheres and self-replenishing plant life for those who have no wish to seek more exotic worlds elsewhere in the Galaxy.

The actual 'building' of planets will present no great complication. It will simply involve assembling sufficient matter at a point on the desired Solar orbit. Great lumps of this matter will be allowed to 'fall' on to each other, as if the Moon were arrested in its orbit and compelled to 'fall' on to the Earth. Gravitational laws predict that when this happens the final object created by the collisions will tend to be spherical, and whenever an eccentricity continues to exist, such as a mountain standing in an unstable position, it will simply collapse. It is believed that all the original planets were formed in this manner.

The huge population centres of a completed Dyson Sphere would consist of millions of different-sized worlds. To an inhabitant of any one of these worlds the daytime sky will seem much like today, and he will see no obvious indication of change. The Sun will be roughly the same distance from him as it is from us. It will appear the same size and it will radiate the same heat. Plants and trees will be as numerous; gravity, if he lives on one of the Earth-sized worlds, will be the same, and day and night will succeed each other at approximately 12-hour intervals. But night will be vastly different. Even during cloud-covered nights, there will be no longer any such thing as pitch blackness. The night sky will be brilliant from the glare of the reflected sunlight of countless celestial objects. So bright will this glare be that only the brightest stars will be visible, and astronomical observatories will have to be moved to the outer reaches of the Solar System.

How will the Sphere be governed, and how will people on one side of the Sun communicate with people on the other? Since they will always remain in the same position relative to each other, and since no radio or television signals can travel directly through the Sun or through the surrounding Solar corona, this difficulty might seem insuperable. Fortunately, this problem has been solved by James Strong, a Fellow of the British Interplanetary Society, who has discovered a means of creating a radio relay system around the Solar System, in such a way that signals can travel between any two points, no matter what obstacles lie between. He has drawn on the work of Joseph Lagrange, the 18th-century French mathematician, who worked out a planetary law which has been called the 'Trojan system'. Lagrange calculated that any object occupying the same orbit as Earth, and forming an equilateral triangle between itself, the Earth and the Sun, will maintain a stable equilibrium. If each angle in the triangle is exactly 60 degrees, then the third object, whether it is a tiny satellite or a full-sized planet, will stay in the same relative position indefinitely.

Lagrange's law was proved in 1906 by the discovery of two groups of 'Trojan' asteroids, named after heroes of Troy, whose fixed positions in relation to Jupiter make two equilateral triangles with the Sun. Strong has shown that it will be easy to communicate with someone on the opposite side of the Sun by passing the signal through a correctly placed Trojan relay satellite.[6] In the Dyson Sphere, the Earth could be one of six Trojan planets surrounding the Sun, each maintaining perfect equilibrium with its neighbour, and acting as governmental and communicating centres. While other worlds in the Sphere would be free to pursue moderately eccentric orbits—provided there were no risks of collisions—the Trojan planets would stay in their places, observing and guarding over all. They will be watching over a community of thousands of billions of human beings who will have one immense source of satisfaction. The Sun's radiation will no longer be wasted. At present, only one part in 500 million of Solar heat and light ever strikes the Earth; the rest pours wastefully away into space. With a Dyson Sphere enclosing the Sun, leaving only its 'top' and 'bottom' clear of all objects, it will be a matter of some pride that at least half of this radiation is at last captured and used for the well-being of mankind.

The people who live in the Dyson Sphere will nevertheless comprise the most conservative and unadventurous of our species. Far from roaming the Galaxy, they will either be living on the Earth or on a world within easy reach of it. As I have argued in

Chapter 10, there are likely to be hundreds of millions of worlds in the Galaxy fit for human occupation. Multitudes of colonial expeditions will set out, and many of these, having found suitable planets, will be under the impression that they have sundered all ties with the growing bureaucratic technology of Earth and its newly-created sister worlds. In some remote stellar systems, they will be justified for some thousands of years in their illusion that they have left the mainstream of history behind them. But what of those who have found comfortable planets in central parts of the Milky Way? They might be compared with a man who has decided to leave town and live in the country, and has bought a cottage in what he imagines is a remote, inaccessible district, only to find that a six-lane motorway is to be built past his back door.

For Dyson Spheres are likely to become as relatively common-place in the Galaxy as motorways have become in modern industrial countries. 'Is there any physical or engineering reason,' asks Dyson, 'why a growing technology should remain confined to the neighbourhood of a single star? The answer seems to me clearly negative.'[7] In an age many thousands of years distant, even the colossal mass of Jupiter will fail to satisfy human demands. Steps will be taken to initiate Kardashev's Phase 3, in which the resources of an entire galaxy are exploited. At first, spheres are likely to be built around nearby stars. It is highly probable that sufficient materials exist, in the sense that most stars have planets. Two planets with masses comparable to Jupiter have been discovered in orbit around Barnard's Star, 6 light-years from the Sun. The star 61 Cygni, 11 light-years distant, is suspected of having a planet with 16 times the mass of Jupiter. Wherever our descendants find it desirable, such planets will be dismantled to form Dyson Spheres. 'If we assume a technology with a strong will to expand,' writes Dyson, 'it will move from star to star in times at most of the order of 1,000 years. It will spread from one end of a galaxy to another in 10 million years, which is still a short time by astronomical standards.'[8]

One difficulty could slow this process, however. A galaxy such as ours contains too much mass in the form of stars, and too little as planets, to be suitable for really efficient and rapid exploitation. There will be only one solution to this problem. Some stars themselves will be dismantled, or blown up, so that the heavy elements inside them can be mined for their raw materials. We can imagine gigantic chunks of iron and the compounds of other heavy elements being hurled out into surrounding space by the artificially-induced supernova explosion of a star. Such an enterprise would not seem in the least fantastic to a Phase 3 civilization, which would be many billions of times richer in power and resources, and deploy billions of times more energy, than even a Phase 2 civilization which has built a single Dyson Sphere. Just as the people of 400 years from now will complain that Jupiter is of no economic value in its present orbit, so their descendants of several thousand years hence will look enviously at some of the stars, particularly those dying, swollen red giant stars which are in the last few million years of their lives, and in whose neighbourhood no habitable planets could exist.

We know a considerable amount about the natural evolution of stars. A star with the mass of the Sun will end its life by swelling into a red giant and then collapsing into a white dwarf. But a star with twice the mass of the Sun will not collapse tamely into a white dwarf. It is liable instead to end its life with a catastrophic explosion, hurling its heavy elements into surrounding space. Now, let us imagine that the people of a Phase 3 civilization, with a much more sophisticated knowledge of stellar evolution than ours, turn their attention to a particular red giant star whose diameter is about 600 times greater than the Sun's. From its present size and luminosity they calculate that its original mass, before it became a red giant, was about two or three times the mass of the Sun. The nature of its final death, therefore, hangs on a knife-edge. It will either collapse quietly into a white dwarf or else it will explode.

Consider what has actually happened. I discussed in Chapter 9 the fate of various stars after they have either collapsed or exploded. But how did they come to collapse or explode in the first place? What tiny factor will make the difference between a quiet collapse and a violent explosion? Our hypothetical star of two or three Solar masses originally consisted almost entirely of hydrogen, which it spent its life fusing into helium. As a result an enormous quantity of inert helium ash has accumulated in the core of the star. This pile of helium ash becomes bigger and bigger as more and more hydrogen is converted into helium.

Gravitational pressures in the star's core become ever stronger. At length the helium ash reaches the critical temperature of about 200 million degrees F., and itself undergoes thermonuclear ignition. This fatal point in a star's life, when helium atoms merge explosively to form atoms of oxygen, carbon and neon, is known appropriately as the 'helium flash'. Tremendous new heat is generated throughout the star. The outer layers swell and expand until the star has filled millions of times its former volume. It remains in this position for several million years. Gravitational pressures then again take over and the whole star begins to collapse upon itself. Now comes the real crisis, for terrific events have been taking place in the star's core. The oxygen, carbon and neon have themselves been accumulating in a pile of ash. This ash-pile itself becomes heated under the gravitational pressure of its own mass. A whole series of thermonuclear transmutations occur, so that at 4,500 million degrees F., the ash is fused into the heaviest elements, such as iron, titanium, chromium, manganese, cobalt, nickel, copper and zinc. The star at this stage is like a gigantic grenade whose pin has been pulled. An explosion in the core of the star is certain. But what happens then depends entirely on the star's original mass. In a star of the Sun's mass, the explosion at the core will not be violent enough to shatter the star's outer layers. The outer layers will, so to speak, damp down and contain the explosion. But in the larger star of two or three solar masses, the material at the core will be so massive that its explosion will rend the outer layers in pieces. Literally in the space of a few seconds, all the layers of the star will be jammed together in a state of shattering collision. All the star's remaining nuclear energy will be released at once. The result is an explosion on a scale of violence that we can scarcely imagine. The heavy elements from the core are hurled out into space. For a few weeks the exploded star shines with the brilliance of 100 million suns.

The problem facing a Phase 3 civilization is to interfere effectively with a dying star at its crisis stage and actually to cause an explosion where, left to itself, the star would merely collapse. For obvious reasons of safety, the target star would have to be situated at least five light-years from any inhabited planet. The explosion would be best managed by unmanned, computer-controlled laser guns in close proximity to the target star. The trick will be to cause large temperature increases at points in the outer layers at the very moment when they are under maximum pressure from the exploding core. This laser-heating method, incidentally, would work on the same principle that physicists are now using in their attempts to create energy in their laboratories by controlled thermonuclear fusion. A star, like the hydrogen in these laboratories, will tend to become unstable if exposed to large amounts of heat from an external source.

The British astronomer Geoffrey Burbidge has suggested that this process of sudden outside interference causing stellar explosions occurs naturally in the centres of some galaxies where stars are very densely packed together.[9] In his theory, the accidental explosion of one star raises the surface temperature of its neighbour, causing it in turn to explode. The result would be a chain reaction which could explain the observed explosions or disruptions of whole galaxies. Whether this *does* happen is an open question, but Burbidge has produced persuasive reasoning that it *could.* Carl Sagan and Iosif Schlovskii, in their fascinating book *Intelligent Life in the Universe,* suggest that a Phase 3 civilization would be capable of building a laser gun that could detonate a supernova.[10] It would operate at gamma ray frequency at a wavelength of about one Angstrom unit. The gun would have an aperture diameter of about 10 yards and an extremely narrow beam width, probably a small fraction of a second of arc. According to Burbidge's theory the flow of gamma radiation at the star's surface would have to generate about 20 billion ergs per second on each square inch of the star's surface, covering an area with a diameter of only 6 miles. The gun would need a power source far greater than the electrical energy now consumed on Earth in a year. But generation of such power would be a minute expense to a Phase 3 civilization. The sudden attack from the laser gun would create a nuclear chain-reaction which would heat up and weaken the star's outer layers at a critical moment. The explosive core would be unleashed. The result would be a colossal explosion which would convulse the whole star. A few years afterwards, when the most dangerous forms of particle radiation from the exploded star were exhausted, the Phase 3 civilization could move in with their magnetic-field generators, and collect the heavy elements in order to build several hundred Dyson Spheres simultaneously.

It is quite possible, of course, that a civilization deploying energy on this scale would find even more efficient ways of exploiting their resources than in building 'conventional' Dyson Spheres. Entire star clusters might be dammed and controlled.[11] Some artificial arrangement of this kind perhaps already exists at the centre of our Galaxy, where many of the stars are much older than our Sun and its neighbours out here on the galactic periphery. Sadly, we have still no means of knowing, since the centre of our Galaxy is obscured from us by great clouds of dust and luminous hydrogen. Some of the billions of other galaxies do indeed show signs of extraordinary disruption. The galaxy Messier 87, in the constellation of Virgo, has shot out into space a titanic flare, a jet of brightness nearly 100,000 light-years in length. It is conceivable (although such an idea is not supported by the slightest evidence) that this flare may be a colossal rocket jet generated by a civilization in Messier 87, who are at work on a project whose nature we could scarcely comprehend. In all probability the jet is the result of a natural explosion, but that is not to say that the natural origin of all cosmic phenomena is self-evident. If the molecular biologists are correct in their belief that life will always evolve in favourable conditions, then it must follow that there are few limits to what man, or any other intelligent creatures, can achieve. We cannot reiterate too strongly the dictum of the physiologist Constantine Generales that although man's physical strength is tiny, he can nevertheless learn to use and control forces quintillions of times stronger than himself.[12] If there is some fundamental law that says that we cannot, during the course of millions of years, occupy and exploit our entire Galaxy of 100 billion suns, then that law is now hidden from us.

14

The God of Spinoza

Man has made great progress since his first ancestors, the blue-green algae, appeared in shallow seas some three billion years ago. Granted the Sun's comparative stability, which enabled life to flourish and develop, it is plausible to suggest that our present technical achievements, and indeed all the future achievements I have predicted, were inevitable from that moment long ago. I do not refer to man's evolution itself; biologists now believe that our rise to supremacy over other creatures was the result of a billion-to-one-chance.[1] But that is a very different thing from saying that the evolution of intelligent minds was an unlikely event. The number of intelligent planetary civilizations in this Galaxy, as I said in Chapter 10, has been estimated by many astronomers at between one million and 10 million. But very few are prepared to suggest that men looking exactly like us are commonplace creatures in the Universe. The evolution of intelligence in some other form seems much more probable. But for chance, the lords of creation might have been apes or dolphins, or more likely some creatures which, because of the strange turns of evolution, have never in fact existed at all. Whatever they might have been or looked like, let us call them quasi-men. They would have needed members that did the work of fingers, advanced sense organs, a high level of individual consciousness, the habit of abstract thought, and some aptitude for mathematics and astronomy. The essential ability to amuse themselves by pondering romantic and fantastic schemes, without at first any hope of their realization, would have developed in a few individuals as progress was made. Minorities among the quasi-men would strive fanatically to realize such schemes—just as human inventors push their ideas forward in the face of bureaucratic opposition.

Such has been humanity's pattern of technological progress, and it is widely supposed that alien communities would advance in a similar way. Science-fiction suggestions of telepathic races in which everybody simultaneously has the same brilliant idea seem very far-fetched. But why does progress take place at all? It does not have to happen. Most terrestrial species never make any progress. An ant-heap today looks exactly as an ant-heap would have looked a million years ago. While most species remain still, *homo sapiens* pushes forward with astonishing rapidity. As the anthropologist Sir Wilfred Le Gros Clark remarks:

> The whole evolutionary history of the primates has been marked by one special feature which obtrudes itself very forcibly on the attention, and that is the progressive expansion and elaboration of the brain. The wile and cunning of the earlier primates have become the intelligence of the higher primates, and man himself has surpassed all other members of the animal kingdom in his capacity for mental activities of the most elaborate kind.[2]

One by one, the secrets of nature are unlocked to his advantage. It all seems too easy. Something about it does not make sense. It may of course be true that our Sun, this medium-sized star on the edge of the Galaxy, has spawned a race of geniuses more intuitive than any since the Galaxy's birth. But any random glance at newspaper headlines somehow discourages this impression. Humanity advances *despite* the petty stupidity of its majority. If nothing, then, is special about humanity, the explanation must lie in nature. Electricity, the atom, orbital mechanics—these and countless other devices of nature have been found available for our use. They might almost have been prepared for us, just as slices of cheese are laid down at strategic points to guide the laboratory rat. It might not be altogether ridiculous to assert that the laws of the Universe have been deliberately programmed to further the advancement of intelligent life. But programmed by whom? We are faced at once with the question of God. Assuming that God—or a supreme being *—does exist and *has* programmed the laws of the Universe, we must ask: what kind of being is He?

* The convenient phrase 'supreme being', implying a creature greater than man but perhaps less than omnipotent, appears first to have been used by Robespierre in a speech on June 8th, 1794, when he set out to discredit the atheist extremists in his own party.

For about ten years, roughly between 1920 and 1930, it was generally admitted that nobody knew more about the cosmos than Albert Einstein. Religious leaders faced the embarrassing fact that he was in a position to know more about their business than they themselves. Not unnaturally, this annoyed and alarmed them. The man of doctrine generally has little love for the man of empirical inquiry. The inevitable clash came. The Jewish magazine *Reflex*, publishing in Chicago in 1929, ran an editorial accusing Einstein of having uttered 'pure blasphemy'. The great man's offence lay in the answer he had given to a rabbi who had asked him, in a blunt, reply-paid telegram, 'Do you believe in God?' Einstein at once cabled back, 'I believe in Spinoza's God who reveals himself in the harmony of all that exists, but not in a God who concerns himself with the fate and action of men.' Pointing out that Benedictus de Spinoza had been excommunicated in 1656 for refusing to believe in guardian angels or immortal souls, the Rabbis declared:

> Spinoza was excommunicated for the very good reason that he denied the personality of God, and tried to chain God to His own laws. Spinoza's God is not free, and if He wanted to direct the destiny of man, He could not possibly do so because He is chained to the immutable laws of nature. This, in the eyes of the pious ones, is worse than pure atheism. It is pure blasphemy.

Some elements of the Catholic Church, also, were appalled at Einstein's apparent relegation of God to a mere series of physical laws. Cardinal O'Connell of Boston preached that Einstein's two theories were 'befogged speculation, producing universal doubt about God and his creation'. An editorial in *L'Osservatore Romano*, which reflected Vatican views, said that O'Connell was right to denounce Einstein's theories because they 'tended to cut off faith in God from human life'. It called his work 'authentic atheism camouflaged as cosmic pantheism'.[3]

It seems impossible that both parties could have been right in this dispute. Even a compromise seems out of the question. The God of the Rabbis and Cardinal O'Connell, the Jewish or Christian God, takes a personal interest in every human being. He either wins them for Heaven or loses them to the Devil. In this theology each person is protected by a guardian angel. But a person *can* repudiate his guardian angel by deliberately turning to sin. He has then, in the orthodox view, endangered his immortal soul by exposing it to seizure by the Devil. A person is free to choose evil, and knows well the difference between good and evil, because his ancestor, Eve, ate the apple from the Tree of Knowledge. The resulting balance of power between God and the Devil permits man to commit great crimes. The two great opponents cannot directly prevent or participate in these crimes. They can only, and very discreetly, encourage or tempt.

This, in a few short sentences, is the 'unified field theory' of the Christian faith. It leaves some questions awkwardly unanswered. Why, for example, must the innocent suffer for the crimes of the wicked? The churchmen answer that physical suffering makes no difference to a man's immortal soul, and that it therefore doesn't matter; the wicked man, on the other hand, becomes even wickeder as a result of his crime, and this is of much greater significance. This answer has always seemed to many people to verge on arrogance. We are asked to believe, without evidence, that an innocent man will enjoy eternal bliss in Heaven after death. In the expectation of such bliss, he ought cheerfully to put up with any horrors inflicted on him during his life. A man might think it pardonable to reflect that the whole cruel system would never have been foisted on him if God had been paying proper attention when Eve ate from the forbidden tree. The whole theory rattles with inconsistencies. According to strictly orthodox interpretations, innocence alone does not qualify a man for admission into Heaven. He must also be devout; he must *believe* in the system if he wants to benefit from it. An innocent but Godless man is still a sinner.[4]

It is as if God were constantly anxious about His own reputation, whereas it might be thought that a truly omnipotent God would not care whether men worshipped him or not so long as they refrained from sin. And where, for that matter, is God at this present time? Is He everywhere in the Universe simultaneously, or can He only be at one place at one time? Arthur C. Clarke has half jokingly suggested that God's movements may be restricted by the speed limit of the Special Theory of Relativity. 'That may be the cause of all our troubles,' says Clarke. 'He's coming as fast as He can, but there's nothing even He can do about that maddening 186,000 miles per second.'[5] Where is He coming *from*, anyway? Perhaps the most brutal comment on Christian theology came from Nikita Khrushchev: 'We sent Yuri Gagarin into space to see if he could find the Kingdom of Heaven. He couldn't see it, so we sent up Gherman Titov to make sure. And he couldn't find it either.' Offensive as this remark may have sounded, Khrushchev was in good company. Saint Augustine is said to have avoided these difficult questions with equal agility. Asked by a scoffer what God was doing before he created the world, he replied sternly, 'He was busy creating Hell for people who ask foolish questions.'

But most of the objections raised here disappear if we believe instead in the God of Einstein and Spinoza, the God who reveals himself in the 'harmony of all that exists'. This God is a much simpler being. He is not in the least interested in individual behaviour—which is just as well, since he lacks all power to control it. His sole interest is the advancement of intelligence, by which I mean technological achievement.* There is some circumstantial evidence that this God exists, or once existed, while belief in the Christian God rests upon faith. This evidence is inadequate for swaying a law-court, but it might qualify as a *prima facie* case. I have already remarked on the curious coincidence that the one inhabited planet in this Solar System should have a Moon out of all proportion to its size. The soft surface of the Moon's marias, like damp sand, might have been made for descending spacecraft. The planets could hardly be in more convenient orbits. All nine of them rotate round the Sun in the same direction, so that a mere 12,000 m.p.h. acceleration is needed to travel from one of their orbits to the next. How very convenient, and how very odd, that Venus, the one planet closest to our own in size, should be so richly endowed with carbon dioxide, the gas that desires nothing more than a diet of algae to turn itself into oxygen! (It may be objected that it would be even more convenient if Venus had been

---

* 'Achievement' is the key word. There is clearly no advantage in technology which only creates ugliness or pollution. As Freeman Dyson once remarked, 'it is easy to see around us examples of technology without intelligence, particularly when we eat lunch at the campus cafeteria at La Jolla and watch the bulldozers demolishing the eucalyptus trees.'

covered already with oxygen and nitrogen, so that no algae treatment would be necessary. But this would be physically impossible unless Venus occupied almost the same orbit as Earth—and its doing so would make catastrophic collisions with our world much more probable over a long period.) The huge planet Jupiter has just the right mass for safe exploitation. It is sufficiently close for human visitation within the technology of the near future. It is massive enough to provide us with a generous supply of raw materials and new worlds, yet it is sufficiently small and distant so that its fragmentation will not disrupt the Earth's orbit. All very curious, we might think.

But the Universe itself possesses three phenomena which should arouse our suspicions even more strongly. Three things make life possible in the Universe. Freeman Dyson calls them the three cosmic 'hang-ups', the term borrowed from psychology meaning something which arrests or delays a normal process. As Dyson explains, 'psychological hang-ups are generally supposed to be bad for us, but cosmological hang-ups are absolutely necessary for our existence'.[6] The first of these is the great distances between celestial objects. When Thomas Henderson found in 1831 by measuring the parallax of the nearest star to the Sun that its distance was no less than 24 trillion miles, or 4·3 light-years, the reaction was general dismay. It seemed that it would be impossible to analyse the substance of objects so distant, let alone visit them. The Solar System appeared a much more lonely place than hitherto.

But the loneliness seems a blessing today when we know of the violent and dangerous things that stars can do. Nova and supernova explosions, which occur periodically in the Galaxy and involve the partial or total destruction of stars, would endanger or destroy any planetary civilization living within two or three light-years of the blast. There would be no damage to buildings, but the radiation fall-out would resemble the aftermath of a major nuclear war. Cities would have to be evacuated, and the population would have to live underground for at least a year.[7] People would emerge after the crisis to find their fields ruined by neglected growth and chaotic genetic mutation. Only in the depths of the oceans would life be unaffected. The effects would be far worse if a supernova occurred within less than one light-year of the Sun. The radiation from the explosion could be energetic enough to trigger a chain-reaction and blow up the Sun. The Earth would then be enveloped in flames and reduced to a cinder.

So we should not be too dismayed by the scale of interstellar distances. Life would be impossible in all but the most remote solar systems if stars were too densely packed together.* The frequent stellar explosions would prevent planets from having the necessary billions of years during which life could evolve free from violent disruption. The interstellar distances, moreover, make the chances of collisions between solar systems extremely remote. Such collisions, even if they did not involve two or more suns crashing into one another, would hurl any planets out of their settled orbits, condemning their inhabitants either to a fiery death or to perpetual frozen night.

The distance between galaxies, on a scale a million times greater than those between stars, is also essential to our safety. One can imagine the danger of living in the vicinity of a quasar, an object which blazes with the brilliance of 100 million suns, or a radio galaxy which spews out enormously energetic electron energy, or a Seyfert galaxy, which appears to be in a state of shattering explosion, caused perhaps by chain reactions of supernovae inside it. Quasars, radio galaxies and Seyfert galaxies are fascinating objects to peer at through telescopes at distances of millions or billions of light-years, but they are the last things we would wish to meet face to face on a dark night.

Distance by itself is not enough to sustain us. The second 'hang-up' is spin. All celestial objects spin at varying speeds. Every stable cosmic structure consists of small masses orbiting around larger masses. Nobody knows where the initial impetus to orbit came from, but the Universe as we know it would cease to exist if it stopped. The Earth would crash into the Sun, and the Sun would crash into the centre of the Galaxy. All the galaxies

---

* Planetary life would probably have little chance to evolve in parts of space where stars are densely packed, like the centres of galaxies, or in the globular clusters where the average distance between stars is about 4 light-months. The surface of planets inside these clusters would be bombarded continuously by cosmic radiation, which would be lethal to any life-form like ours.

would crash into all the others, and there would be a swift return to the condition of the 'primeval atom' from which the Universe probably began. This is likely to happen anyway in about 70 billion years' time, according to current belief, when all orbits will have decayed and when the expansion of the Universe has long exhausted itself. But I doubt if any reader of this book is seriously alarmed by the thought of what will happen in 70 billion years' time. The point is that such a huge period of time, when added to the 12 billion years or so which have already passed since the birth of the Universe, gives countless civilizations the chance to be born, to flourish and to die on innumerable worlds circling around placidly shining stars.

This brings us to the third and perhaps the most spectacular 'hang-up', that of thermonuclear fusion. About 90 per cent of the Universe consists of hydrogen, and hydrogen has the agreeable habit of 'burning' slowly and quietly into helium. This process of fusion takes billions of years in an average-sized star like the Sun, while it fills the space around it with a constant and life-giving radiation. It may be wondered, if the Sun is nothing more than a gigantic hydrogen bomb, why it does not explode as our terrestrial models do. The reason is that the Sun burns 'pure' or light hydrogen, whereas to construct our deadly bombs we use 'heavy' hydrogen with the heavy hydrogen isotopes known as deuterium and tritium. The fusion of light hydrogen with its weak reactions takes place about a quadrillion times slower than the 'strong' reactions of heavy hydrogen. This is very fortunate, since if stars habitually blew up the moment their fusion processes began, the galaxies would have been destroyed even before they had been formed, and planetary life would be impossible. Moreover, as one writer has remarked, the oceans, with their deuterium content, would be excellent sources of thermonuclear explosives, and would continually tempt would-be builders of 'doomsday machines'.[8]

But when we examine our supply of raw materials and their availability here on Earth, the jigsaw pieces seem to click into place with uncanny timing and precision. Coal, oil, fissionable uranium, and countless other precious raw materials are found in the Earth's crust and in the oceans in such abundance as almost to raise the suspicion of some cosmic design. It is astonishing, in any case, that these commodities have been able to survive perhaps 100 million years of violent geological and atmospheric change. Coal and oil have been buried at shallow depths beneath the surface, and thereby saved from oxidation, by processes that we are far from understanding. They were found to be easily accessible in the 18th and 19th centuries and even earlier, when our technology was primitive, and when we knew nothing of the uses of uranium. Uranium itself has survived the Earth's long history almost as if it had been waiting for us to exploit it. The atomic nucleus of fissionable uranium presents us with another interesting 'hang-up'. It may be compared with a raindrop which is kept spherical by the surface tension of its water molecules. But the difference is that uranium's surface tension is trillions of times stronger than water's. As a result, less than one part in a million of the Earth's uranium has been dissipated since the world was formed.

But these fuels exist only in very limited quantities. Those natural energy resources of the planet which have so far been exploited cannot be expected to last more than another 100 years or so. Discoveries of new oil fields even on the scale of those in Alaska and on Britain's continental shelf cannot indefinitely satisfy our demands for energy. Natural gas and uranium must come to an end. Many other precious metals important to industrial nations will in time run out. But it does not appear that this will matter. In the longer term, we have on Earth almost unlimited deuterium for our energy and a growing plastics technology for our industrial raw materials. We have Solar energy in the still longer term, with all the benefits to come from mining the Moon and exploiting the asteroids and planets. And, in the thousands of years to come, we shall exploit the energies to be found in hundreds of thousands of stars. Even the terrific but still mysterious energy sources of pulsars and quasars may be discovered and put to use. A group of the world's foremost physicists has informed the American Government that this should one day be possible.[9] Today's pure science often becomes tomorrow's industrial technology. Francis Bacon enshrined this principle in the famous Latin tag *Natura non vincitur nisi parendo*—Man can only conquer nature by obeying her.

Looking into the past to the remote days when fire was first used for cooking and warmth, we can draw a line to the present which records the history of an ever-improving technology. I have tried to show in this book that it is perfectly reasonable to extrapolate this line into the future and postulate that our activities may be expanded into the Galaxy. In the space perhaps of 20,000 years we shall have traversed the road from caves and wattle huts to the Galactic Imperium. Yet all this will have been achieved despite the stupidity and conservatism of the majority of people in nearly all ages. It will seem almost unbelievable to our descendants, for instance, that the year 1972, when breakthroughs of every kind were being made in space and planetary exploration, should have been marked by an inter-governmental conference in Stockholm attended by delegates from 112 nations with the official slogan of 'Only One Earth'.

Yet we have always had a tendency to think in this reactionary way. The majority of mankind simply fails to see the significance of what the scientific minority is doing. Through some biological quirk which distinguishes us from other species, we have nearly always had practical and energetic minorities who make the scientific progress that carries forward the rest. Their work is often despised and ignored—until its benefits become apparent, and everyone steps forward to claim them. Nobody knows why human beings, alone among known species, should have the gift of a capacity for original thought. But that gift will take us to the stars.

Many radical would-be reformers advocate an 'alternative technology', in which we would take our energy from such 'natural' sources as windmills and watermills or the heat from the interior of the Earth. Engineering activities would then be restricted, pollution problems would be solved, and the world (in their view) would be a happier place.[10] One might also add that the human mind would stagnate under these restraints, and that a few generations of such rigid government would rob us of our creativity and make us essentially *un-human*. These ideas seem ideologically related to those Rousseau-like dreams, which are still occasionally touted, of a 'return to nature' and the desirability of living without technical aids like the 'noble savage'. We hardly need Robert Ardrey to tell us that the savagery of Neanderthal and Cro-Magnon Man greatly exceeded his nobility.

The one cardinal error which these reformers make is to assume that man is free to choose his own long-term future. He is indeed free to choose it within narrow limits. Like a snake compelled to move through a tunnel, he may move from side to side, experimenting with this way of life or that. But there is only one forward path, the path to unending technological expansion. The present example of Communist China ought to be a lesson to these thinkers. For many years Mao's government experimented with an almost wholly agricultural economy. China was held up to us as proof that man could get along without high technology. Then, at the end of the 1960s, the Chinese decided to chuck the whole idea. They at last found it intolerable that Japan's economy should so outstrip their own. They are now buying fleets of civilian jet aircraft, entering the computer business, launching scientific space satellites, and in the consumer field they are manufacturing high-quality cameras at prices intended to undercut the Nikon and the Pentax.[11] Like many another ambitious nation, the Chinese have found that a modern civilization without machines is like a man without limbs.*

The long-term trend towards ever greater mechanization was given a new and powerful impetus by Francis Bacon. In the 20th century, two important rulers have tried to rebel against Bacon's philosophy. Both failed. The first of these was Hitler, who drove genuine scientists from his country and proclaimed that 'a new and magical interpretation of the world is coming'.[12] The second, Mao Tse-tung, had a more flexible mind and was able to change his policy when he saw that it was leading his country to ruin. All such 'rebellions' in the future, even if at first appearing to succeed, will be similarly doomed. All the passionate ideological movements which can be imagined, all possible brands of millenarianism, Communism, militarism, anarchy or Fascism, will avail themselves nothing against the dead hand of Francis Bacon. Let these enemies of human progress exercise their free will to the full. Let them attack by propaganda or actual subversion, by local acts of terror or wholesale devastation, by single bombs or a holocaust,

---

* The pasha in Kinglake's *Eothen* seems to have understood this plainly when he exclaimed in admiration, 'The armies of the English ride upon the vapours of boiling cauldrons, and their horses are flaming coals. Whirr! Whirr! All by wheels. Whiz! Whiz! All by Steam.'

by declared war or sudden ambush, and they will always lose. The Baconian scheme can be delayed, but it cannot be stopped. People who think otherwise, who seek to relegate large parts of humanity to the status of animals, and who thereby challenge Bacon's law of progress, take the risk of being answered, as the Japanese warlords were answered at Nagasaki and Hiroshima, by a Power that admits of no dispute.[13]

Useful things will continue to be invented. And the very continuation of such inventions is a justification for faith in some great design. The treasures lying in our path, and our human capacity to use them, may for all we know have been prepared by some cosmic intelligence with an interest in our advancement. It has seemed worthwhile to speculate on the ultimate reasons why our development is, in general, so inexorable and so rapid. Whatever setbacks occur, we shall go forward to ever more magnificent achievements. I have suggested what some of these achievements may be, although the fulfilment will no doubt be even more spectacular. The events predicted in this book, or something like them, will sooner or later come about. Because of short-term uncertainties, we cannot tell *when* they will happen. But we can say with absolute surety that they will happen.

A Do-it-yourself Guide to the Special Theory of Relativity

The Special Theory, which forms the basis for both the General Theory of Relativity and the laws of geometrodynamics, rests on the fact that the speed of light in the vacuum of space is constant at 670 million m.p.h. (or 186,000 miles per second, or 300,000 kilometres per second), irrespective of the speed of its source. Mathematical logic therefore dictates that the *length* of a moving spacecraft in the direction of motion must be reduced in proportion to its speed. The spacecraft's *mass*, i.e. the energy required to accelerate it, is increased as the object accelerates. And all clocks aboard the spacecraft—the term 'clock' here includes both the astronaut's wristwatch and the ageing process of his body—slow down as the spacecraft accelerates.

The extent of these changes are calculated by three simple formulae. To calculate the *length* of a moving spacecraft, we multiply its length at rest, that is its length when stationary on a planet, by the formula

$$\sqrt{\left(1 - \frac{v^2}{c^2}\right)}$$

where v is the speed of the spacecraft, and c is the speed of light. The increased mass of the moving spacecraft is calculated by a slightly different formula. Its mass at rest is multiplied by

$$\frac{1}{\sqrt{\left(1 - \frac{v^2}{c^2}\right)}}$$

Measuring the slowing of time on board the spacecraft is equally simple. To see how much more slowly the astronauts are ageing, we multiply a given period, say 60 minutes of Earth-time, by

$$\sqrt{\left(1 - \frac{v^2}{c^2}\right)}$$

Two spaceships cannot recede from one another at a combined speed greater than that of light. If a man standing on the ground sees one craft racing overhead going due north at 90 per cent of the speed of light, and another going due south at the same speed, he might suppose that each craft was receding from the other at a combined speed of 180 per cent of that of light. Yet he would be wrong. The sum of the two speeds cannot exceed that of light. It must be calculated by the formula

$$\frac{a + b}{1 + \frac{ab}{c^2}}$$

where a and b are the respective speeds of the two craft, and c is the speed of light. It will be seen from this formula that if the two craft had been travelling very slowly, say 600 m.p.h., then the sum of their speeds would be about 1,199·9999999 m.p.h., or *almost* 1,200 m.p.h. But if their speeds are very great, the formula gives a quite different kind of answer. Pretend that instead of two spaceships the man sees two light-beams receding in opposite directions. He will estimate their mutual speed of recession as twice the speed of light or 2c. But if he were riding on one of the beams, he would estimate the other's speed of recession, according to the formula, as

$$\frac{c + c}{1 + \frac{c^2}{c^2}}$$

which of course works out as c.

Let us therefore test an imaginary spacecraft at varying speeds, and now see how its length, ageing and mass change according to Einstein's equations. The slowing of time, in particular, will make interstellar journeys possible within reasonable ship-time if journeys through Superspace prove impossible. We will assume that a ship, while stationary in port, is exactly 100 yards in length, and has a rest-mass of 100 tons. As the ship accelerates we see that a ship-hour becomes a progressively smaller fraction of an earth-hour.

Speed of ship as percentage of light	Length of ship (yards)	Mass (tons)	Duration of ship-hour in minutes (Earth = 60)
0	100·00	100·00	60·00
10	99·50	100·50	59·52
20	97·98	102·10	58·70
30	95·39	104·83	57·20
40	91·65	109·11	55·00
50	86·60	115·47	52·10
60	80·00	125·00	48·00
70	71·41	140·03	42·75
80	60·00	166·67	36·00
90	43·59	229·42	26·18
95	31·22	320·26	18·71
99	14·11	708·88	8·53
99·9	4·47	2,236·63	2·78
99·997	0·71	14,142·20	1·17
100	zero	infinity	zero

As these figures plainly indicate, no spacecraft can ever travel at the speed of light itself. An infinite mass would require an engine of infinite power for its propulsion. And even if this were miraculously achieved, the spacecraft's length would be zero, and it could not therefore exist.

APPENDIX II

Dimensions of the Solar System

1. *The Planets*

Planet	Average distance from sun (millions of miles)	Diameter (miles)	Mass (Earth = 1)	Density (water = 1)	Weight of 150-lb man at surface (lbs)
Mercury	36	3,000	0·05	5·5	55
Venus	67	7,600	0·81	5·3	130
Earth	93	7,900	1·00	5·5	150
Mars	142	4,200	0·11	3·9	55
Jupiter	483	89,000	317·80	1·3	380
Saturn	886	75,000	95·20	0·7	160
Uranus	1,800	30,000	14·50	1·7	155
Neptune	2,800	31,000	17·20	1·8	210
Pluto	3,700	3,600 (?)	0·80	unknown	unknown

2. *The Moons*

Planet	Moon, in order of distance from planet	Average distance from planet (miles)	Diameter (miles)
Mercury	None known		
Venus	None known		
Earth	The Moon	240,000	2,160
Mars	Phobos	5,800	12(?)
,,	Deimos	15,000	7(?
Jupiter	Amaltheia	110,000	70
,,	Io	260,000	2,100
,,	Europa	420,000	1,800
,,	Ganymede	670,000	3,100
,,	Callisto	1,200,000	2,800
,,	Hestia	7,100,000	50
,,	Hera	7,300,000	20
,,	Demeter	7,400,000	10

Planet	Moon, in order of distance from planet	Average distance from planet (miles)	Diameter (miles)
Jupiter	Adrastea	13,000,000	10
,,	Pan	14,000,000	10
,,	Poseidon	14,600,000	10
,,	Hades	14,700,000	10
Saturn	Janus	98,000	190
,,	Mimas	120,000	300
,,	Enceladus	150,000	400
,,	Tethys	180,000	560
,,	Dione	240,000	430
,,	Rhea	330,000	840
,,	Titan	760,000	3,080
,,	Hyperion	920,000	100
,,	Japetus	2,200,000	830
,,	Phoebe	8,100,000	100
Uranus	Miranda	77,000	200
,,	Ariel	120,000	500
,,	Umbriel	170,000	350
,,	Titania	270,000	600
,,	Oberon	360,000	500
Neptune	Triton	220,000	2,300
,,	Nereid	3,500,000	200
Pluto	None known		

3. *The Asteroids, or minor Planets: the largest 10*

Asteroid	Average distance from Sun (millions of miles)	Diameter (miles)
Ceres	257	427
Vesta	219	370
Pallas	257	280
Hygeia	293	220
Juno	248	150
Metis	222	135
Astraea	239	111
Hebe	225	106
Iris	221	93
Flora	204	77

Aspects of some problems concerned with the Construction of Dyson Spheres

IAIN K. M. NICOLSON

Lecturer in Astronomy at The Hatfield Polytechnic Observatory

Dyson has suggested[1] that advanced technological civilizations elsewhere in our Galaxy—if any such exist—would wish to utilize the entire energy resources of their parent stars by surrounding these stars by spheres designed to collect all the radiation being emitted from them. The energy collected by these spheres would be available to perform work, and would be re-radiated from the outer surface of such spheres largely in the form of infra-red radiation. Dyson demonstrated that it should be possible to build such a 'Dyson Sphere' in the Solar System by dismantling a planet and rearranging its material.[2]

The Dyson Sphere as originally envisaged could be a fairly lightweight structure. Adrian Berry suggests that the construction of such a Sphere in our system will eventually be undertaken, and that Earth-sized planets could be constructed as part of the Sphere, which would be furnished with atmospheres and made habitable for man.

The amount of matter necessary to construct (a) a basic Dyson Sphere with the minimum mass necessary to collect and utilize solar energy, and (b) a sphere containing many Earth-sized planets, is naturally very different. Dyson has shown[3] that large, lightweight, rigid structures can be constructed in space and that the Sun could, for example, be surrounded by a system of octa-hedral structures of size 10^6 km. whose total mass would be only 10^{-5} Earth-masses. However, these structures would be essentially open frameworks and thus full of holes. In order to intercept all the available sunlight, the 'holes' would have to be filled in, and the mass of the sphere considerably increased. However, if mass equivalent to the Earth's were arranged in the form of a spherical shell round the Sun, the radius of the shell being the radius of the Earth's orbit, then about 3 grams of material would be available per square centimetre of surface area. This would be perfectly adequate for the construction of a system of structures to collect and utilize solar radiation. The planet Venus contains sufficient matter for this purpose.

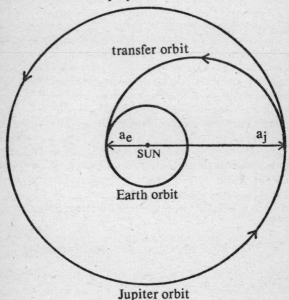

Fig. 1

A sphere of type (b) could be constructed only by dismantling one of the giant planets, in particular Jupiter which has a mass roughly 318 times that of the Earth.

Let us consider the consequences for the Earth of the demolition and reconstruction of Jupiter. Several questions arise, the most obvious being whether the destruction of Jupiter would so perturb the orbit of the Earth as to produce drastic climatic changes, trigger ice ages, and so forth. It seems quite certain that no such drastic consequences would ensue. The effect of Jupiter on the Earth at present is a very small perturbation of the orbital elements. Jupiter is largely responsible for the slow change in the orientation of the Earth's orbital plane. The maximum gravitational force directly exerted on the Earth by Jupiter is never more than 1/16,000th of the solar attractive force. Even the total removal of Jupiter's mass from the Solar System would have no appreciable consequences for the Earth.

If Jupiter's mass were rearranged into a spherical shell of uniform density outside the Earth's orbit, as envisaged by Dyson, what effect would this have on it? It is a standard result in classical mechanics[4] that the net gravitational attraction on a body lying within such a shell or, indeed, within a coplanar ring, is zero. Thus, the net gravitational attraction of the Dyson Sphere on the Earth, provided that the Earth were contained within it, would be zero, and no perturbations of the orbit would result.

To estimate the amount of energy required to transport the mass of Jupiter to the vicinity of the Earth's orbit and to establish it in a similar orbit to that of the Earth, it is assumed that the minimum quantity of energy will be required if Jupiter's mass is transported along an ellipse whose perihelion distance is the radius of the Earth's orbit (see fig. 1).

The energy of a particle of unit mass moving in Jupiter's orbit, E_j, is given by

$$E_j = \frac{-GM_o}{2a_j}$$

where M_o is the mass of the Sun, a_j is the semi-major axis of Jupiter's orbit, and G is Newton's gravitational constant of 6.67×10^{-8}. The semi-major axis of the transfer orbit from Jupiter to the Earth is $\frac{1}{2}(a_j + a_e)$ where a_e is the semi-major axis of the Earth's orbit, and hence the energy of a particle in the transfer ellipse is

$$E_t = \frac{-GM_o}{(a_j + a_e)}$$

Since the energy of a particle in Earth orbit is

$$E_e = \frac{-GM_o}{2a_e}$$

we then have the total energy required to move Jupiter's mass along such a path given by

$$E = -GM_oM_j\left(\frac{1}{2a_j} - \frac{1}{2a_e}\right)$$

Putting in values for these quantities, we obtain

$$E = 10^{43} \text{ ergs}$$

as a minimum value for the energy required.

We shall now consider some possible methods of dismantling planets in general and Jupiter in particular.

1. *Disruption by explosion.* In principle, a planet could be blown up by some as yet unconceived explosive. One explosion of sufficient violence might suffice, or the operation might be accomplished by means of a celestial mining operation consisting of a series of blasts.

2. *Centrifugal disruption.* Dyson has suggested[5] that conductive windings be laid round a planet, giving rise to magnetic fields which would produce a torque upon the planet. Consequently, a planet's rotation could be accelerated to the point where disruption would take place. Alternatively, a civilization which has mastered fusion power might be able to use reaction thrust to accelerate a planet's rotation, with the same result.

3. *Fusion and redistribution of the gas giants.* The giant planets Jupiter, Saturn, Uranus and Neptune, are primarily gaseous bodies of which the principal constituents are hydrogen, helium and hydrogen compounds. J. H. Fremlin and Anthony Michaelis have suggested[6] that, given mastery of fusion techniques, suborbital fusion and transmutation reactor satellites might be inserted into the atmosphere of, say, Jupiter, which would suck in hydrogen and convert it first into helium and then successively into heavier elements such as iron. The output of these processes would then be placed into orbit round Jupiter and redistributed to form a Dyson Sphere.

Let us now consider the relative merits of these schemes for the demolition of Jupiter.

Method 1: Disruption by Explosion

The total amount of energy required to disrupt a planet completely is of the order of the gravitational potential energy of the planet, i.e. not less than

$$\frac{GM_j{}^2}{2r_j}$$

where G is the gravitational constant, M_j is the mass of Jupiter, 2×10^{30}g., and r_j is the radius of Jupiter, $7\cdot1 \times 10^9$cm (equatorial value). On this basis the energy required is of the order 2×10^{43} ergs. As the Sun radiates 4×10^{33} ergs per second, this is equivalent to the total energy radiated by the Sun in about 160 years.

By way of comparison with present-day explosives, this energy requirement is equivalent to about 10^{21} hydrogen bombs each of 40 megatons. Even if it were possible to devise a method of total destruction of matter it would require the conversion into energy of 2×10^{23}g., the mass of a fair-sized minor planet.

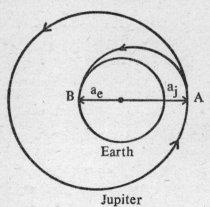

Fig. 2

An obvious criterion is that no fragments of such an explosion should intersect the Earth's orbit. This sets a limit on the velocity at which fragments may leave the vicinity of Jupiter, which may be obtained by considering the minimum velocity a particle would need to reach the Earth from Jupiter. A particle moving along an elliptical trajectory which just touches the orbits of Jupiter and the Earth will, at A in fig. 2, have velocity given by

$$v^2 = GM_o\left(\frac{2}{a_j} - \frac{1}{\frac{1}{2}(a_j + a_e)}\right)$$

therefore $v = 7\cdot5$ km/sec.

Now, the velocity of Jupiter in its orbit is about $13\cdot5$ km/sec. Therefore a particle which is going to move along the path AB from Jupiter will have to be ejected so that its velocity after 'escape' from Jupiter is 6 km/sec. in the direction opposite to the orbital motion of Jupiter. Its velocity relative to the Sun will then be $(13\cdot5 - 6) = 7\cdot5$ km/sec. as required. This indicates that no fragments from the disruption of Jupiter should have ultimate velocities relative to the centre of mass of the explosion greater than 6 km/sec.

Furthermore, particles having this ultimate velocity (relative to the centre of mass of the explosion) in the same direction as Jupiter's present motion would exceed the escape velocity of the Sun, and be lost from the Solar System.

It seems most unlikely that an explosion could be engineered satisfying these criteria, even if a suitable explosive were available.

A most telling objection to the explosive method lies in the composition of Jupiter itself. It is not known whether or not the planet has a core of heavy elements, but most current models of the planet suggest that such a core will be small or totally absent. The composition of Jupiter, according to the model of DeMarcus, is approximately 78 per cent hydrogen, and the remainder largely helium. Assuming a similar abundance of heavy elements in this composition to that of the Sun, the proportion of heavy elements in the planet, if no metallic core exists, will be about one to two per cent, i.e. 3 to 6 Earth masses. Since only the heavier elements can be used for constructional purposes, the explosive disruption of Jupiter would be extremely wasteful, yielding only a few Earth masses of usable material.

Method 2: Centrifugal disruption

The use of magnetic forces to produce a torque acting to accelerate a planet's speed of rotation to the point where the equator
takes off into space has been fully described by Dyson.[7] In the
case of the Earth, the net torque so produced would be

$$T = \frac{4}{15} R^3 H_o^2$$

where R is the radius of the Earth, and H_o the induced magnetic
field.

The angular acceleration of the Earth so produced is given by

$$A = \frac{T}{I_e}$$

where I_e is the moment of inertia of the Earth, $9\cdot7 \times 10^{44}$ g./cm².
If 100 Gauss is taken as a realistic value of H_o, then $A = 9 \times 10^{-16}$
radians/sec. After 40,000 years of continuous application the
rotation rate of the Earth would reach 10^{-3} radians/sec., and the
equator would disrupt. If this method were applied to Jupiter,
again assuming H_o of the order 100 Gauss, then since the
moment of inertia of Jupiter, $I_j = 3\cdot9 \times 10^{49}$g.cm², the angular
acceleration would be

$$A = \frac{T}{I_j} = 2 \times 10^{-17} \text{ radians/sec.}$$

The present rotation rate of Jupiter is $1\cdot8 \times 10^{-4}$ radians/sec.
and the rate at which disruption of the equator will take place is
6×10^{-4} radians/sec. With this value of A, the time taken to
reach the required rate would be about 6×10^5 years. The power
required for this operation is approximately torque, T, times
angular velocity W:

$$TW = 5 \times 10^{28} \text{ ergs/sec.}$$

This is equivalent to $1\cdot25 \times 10^{-5} \times$ total solar power, or about
6,000 times the solar power intercepted by Jupiter. If solar power
were to be used to generate the current required for the accelerating magnetic field, a structure of at least 80 Jupiter radii in radius
would be required to collect the radiation.

In principle, as an alternative to magnetic acceleration, re-action thrust—the rocket principle—could be applied in the form of a large number of rocket motors drawing power from, say, fusion sources. Although this technique could readily be applied to a planet with a solid surface, such as Mercury, Venus, Earth or Mars, the engineering difficulties would be more acute in a gaseous body like Jupiter.

As in the case of Method 1, if this technique could be applied successfully to Jupiter, only a few Earth masses of useful material would be made available for constructional purposes.

Method 3: Fusion and redistribution of Jupiter's mass

If the techniques of controlled fusion can be mastered, and evolved to the stage where heavy elements may be synthesized by fusion (i.e. if the processes going on in massive stars can be duplicated artificially), then in principle Jupiter could be converted from useless light elements to useful heavy elements such as iron. Advanced 'reactors' might be constructed which could be inserted into Jupiter's atmosphere to perform this task, the heavy elements produced being ejected into Jupiter orbit and then moved to their required destinations, probably by fusion power since plenty of energy would be available from the process itself. In this way the entire mass of Jupiter, 318 Earth masses, could be made available for the construction of the form of Dyson Sphere envisaged by Adrian Berry.

The fusion reactors might be powered initially by solar power collected by a structure in Jupiter's orbit, then—glossing over the engineering problems—the excess energy produced by the process could be utilized to power further reactors, and so on.

If such a process could be accomplished, the major problem would be an excess of energy produced as a result of the reactions.

Fusion reactions converting hydrogen to iron release something like $7 \cdot 6 \times 10^{18}$ ergs per g of hydrogen. The mass of Jupiter is about 2×10^{30}g, and thus the total amount of energy released as a result of converting Jupiter into heavy elements is of the order of $1 \cdot 5 \times 10^{49}$ ergs.

This is equivalent to the energy produced by the Sun in about 10^8 years. If the reactions were able to proceed at the same rate as the Sun was producing energy, then it would take about 3×10^5 years to produce material equivalent to the mass of the Earth, or about 4,000 years to produce the mass of the Moon. If all the radiation produced in the reactions under these conditions were radiated isotropically, the Earth would intercept from the direction of Jupiter a quantity of radiation equivalent to 4 per cent of the radiation presently received from the Sun, possibly in the form of dangerous short-wave radiation.

But if the problem of protecting the Earth could be overcome, then the rate of dismantling and converting Jupiter into heavy elements would be governed by the rate at which the reactors could work, and the timescale cut appropriately.

It is difficult at this time to envisage how much of the energy produced in this way could be utilized and how much would be wastefully radiated away. For example, we have seen that the energy required to move Jupiter's mass to the vicinity of the Earth is about 10^{43} ergs. If we assume that the entire task of dismantling Jupiter, moving the fragments to their desired locations and then reconstructing them requires a total of 10^{44} ergs, then this is only 10^{-5} of the energy released in the process. Running the reactor would use part of the energy produced, but it is difficult to avoid the conclusion that much energy would be wasted unless a 'mini-Dyson Sphere' were constructed around the planet. This could, perhaps, serve the function of a celestial industrial estate.

Economics

Clearly this type of operation would require vast economic resources on a scale which dwarfs to insignificance the cost of any major project such as the Apollo programme. However, as Dyson has indicated,[8] the economic growth necessary is of the order of a factor of 10^{10}. Even if a fairly conservative growth rate of 1 per cent per annum is assumed, this would result in a doubling of economic resources in about 75 years and a 10^{10} increase in about 25,000 years.*

A 4 per cent annual growth-rate (which can certainly be

* It is worth noting that to a civilization with economic resources of this magnitude, the equivalent cost of the entire Apollo programme in present-day terms would be about £1.20, or $2.40.

achieved over a short period) would result in a doubling of economic resources in about 17 years and a 10^{10} increase in less than 600 years.

Assuming continuing exponential economic growth, it therefore seems quite possible that the economic resources necessary to begin to construct a Dyson Sphere could be available within a thousand years.

Conclusions

Technically, the first stage might be to construct a lightweight Dyson Sphere by dismantling one of the terrestrial planets, preferably Mercury, which, being too hot to be habitable, will be of little economic value. This could be done by the magnetic acceleration method suggested by Dyson. Mercury's mass could be combined with the largest moons of Jupiter and Saturn, which would be dismantled by the same means.

It may then be possible to use intercepted solar energy to power the first reactors in the Jovian atmosphere. These would then supply energy from the fusion process to run other reactors, steadily converting Jupiter into heavy elements which could then be used to construct the massive bodies envisaged by Adrian Berry. The excess energy produced would be useful for various other technological projects. However, if the energy requirements of an expanding civilization continue to grow exponentially, it might be possible to gear the rate of conversion of Jupiter, and hence the production of excess energy, to the needs of that civilization so that comparatively little was wasted. A sphere containing massive planets, possibly endowed with atmospheres, requires the maximum possible usable quantity of material. Jupiter is the most massive planet in the Solar System, and would have to be dismantled. Disruption of Jupiter in its present form, by explosion or by acceleration to cause progressive disintegration, would only release a few Earth masses in the form of useful heavy elements as the planet consists almost entirely of hydrogen and helium. The only method which will release the entire mass of Jupiter for constructional purposes is the application of fusion to convert hydrogen to heavy elements.

If the fusion process can be made to work on a sufficiently large scale to make the method feasible, then ample energy is available to rearrange the Solar System. However, the Earth will have to be protected from the waste energy radiated away; even if it takes as long as 10^8 years to complete the final operations, the energy output rate from the process is the same as the rate of energy generation in the Sun. A lightweight sphere should first be constructed from the fragments of a dismantled Mercury and the giant moons to give the Earth this protection. The excess energy produced by the dismantling of Jupiter could still be useful for technological projects of various kinds.

If the technological problems (including the utilization of excess energy) can be overcome on a similar timescale to the rate of economic growth, then one must admit the possibility that a Dyson Sphere could be constructed.

Glossary

The following may assist some readers even if it insults the intelligence of others.

Asteroid A tiny planet, usually less than 400 miles in diameter.

Billion Throughout this book, I have used American billions, i.e. thousands of millions. A trillion is therefore a million million.

Black hole A region in space where a star vanishes altogether out of this Universe.

Constellation Any one of 88 areas of the sky, arbitrarily mapped out for astronomical convenience.

Cosmology The study of the shape, size, origin, future, and peculiarities of our Universe, and/or its possible relationship to other universes.

Erg The basic unit of energy. I have invented the term 'ergometrics' to describe the science which relates the deployment of energy to the growth of a civilization.

Escape velocity The speed at which an object must travel to escape from a planet's gravitation. The Earth's escape velocity is about 25,000 m.p.h. (40,000 k.p.h.).

Galaxy A large group of stars. The Galaxy to which our Sun belongs contains about 100 billion Suns. It is about 100,000 light-years from end to end. There are tens of billions of galaxies in our Universe.

Geon A particle consisting of pure gravitational energy, but which is nevertheless endowed with its own mass.

Gross national product The total wealth produced and deployed by a nation, after allowing for inflation. It can be expressed in ergs as well as in money units.

Interplanetary space The space between the planets.

Interstellar space The space between the stars.

Light-year A common unit of distance in astronomy. It is the distance which light travels in one year, going at 670 million m.p.h. (1·1 billion k.p.h.), or 186,000 miles per second. It is thus about 5·9 trillion statute miles, or 9·5 trillion kilometres.

Main-sequence star A star in the stable period of its life.

Nova An explosion which partly disrupts a star, after which the star returns to normal.

Photon A particle of light.

Photosynthesis The process by which sunlight nourishes plant life by forming organic compounds from water and carbon dioxide.

Pulsar, or Neutron Star. The remains of any star that was once 50 per cent more massive than the Sun. A pulsar is only a few miles across, but is so dense that material from it the size of a sugar cube would weigh 100 million tons. (See also White Dwarf.)

Quasar Mysterious objects, apparently at the very edge of our Universe, which are estimated to be shining each with the brilliance of 100 galaxies.

Solar System or stellar system. The region occupied by a sun and its planets.

Star A sun, of which the next nearest to our own is more than 4 light-years distant.

Supernova The total disruption of a star by violent explosion.

Superspace The overall universe, of which our Universe is but a part. Known physical laws break down in Superspace, and 'smoke comes out of the computer'.

Universe This word has undergone a subtle change of meaning in the last 15 years. It used to mean the totality of planets, stars and galaxies that exist, and the volume of space they occupy. But now we sometimes describe all this with the term 'our Universe', being careful not to preclude the existence of others.

White dwarf The remains of a star that once had the mass of the Sun. About the size of the Earth, it is very dense, although not nearly so dense as a Pulsar. Material from it of sugar cube-size would weigh about 5 tons.

White hole The opposite of a Black Hole. It is a region in space at which, in theory, matter emerges from another universe or even from a widely separated point in our Universe.

Bibliography

(In the case of two or more authors, the book is listed under whichever is first in alphabetical order. Several titles have been added since the 1974 editions.)

ADAMS, Carsbie C., Frederick I. Ordway and Mitchell R. Sharpe, *Dividends from Space* (Thomas Y. Crowell, New York, 1971).

ALDISS, Brian W., and Harry Harrison (eds), *Farewell, Fantastic Venus!* (Macdonald, London, 1968; Panther Books, London, 1971).

AMERICAN ASTRONAUTICAL SOCIETY, *Space Age in Fiscal 2001* (Vol. 10 of A.A.S. Science and Technology Series, 1967).

—— *Voyage to the Planets* (Vol. 16 of A.A.S. Science and Technology Series, 1968).

ANDERSON, Poul, *The Infinite Voyage* (Collier-Macmillan, London, 1969).

ARDREY, Robert, *The Territorial Imperative* (Collins, London, 1967).

ASIMOV, Isaac, *Asimov's Biographical Encyclopedia of Science and Technology* (Doubleday, New York, 1964).

—— *Photosynthesis* (Allen and Unwin, London, 1970).

—— *Jupiter, The Largest Planet* (Lothrop, Lee and Shepard, New York, 1973).

ASIMOV, Isaac, with Robert McCall, *Our World in Space* (Patrick Stephens, Cambridge, England, 1974).

BACON, Sir Francis, *The New Atlantis*, edited by Harold Osborne (University Tutorial Press, London, 1937). Also, Vol. 3 of *The Works of Sir Francis Bacon*, edited by J. Spedding, R. L. Ellis and D. D. Heath (Longman, London, 1857).

BAXTER, Raymond, and James Burke, edited by Michael Latham, *Tomorrow's World* (B.B.C. Publications, London, 1971).

BECKERMAN, Wilfred, *In Defence of Economic Growth* (Jonathan Cape, London, 1974).

BERGAMINI, David, *The Universe* (Life Science Library, 1962 and 1969).

BERGIER, Jacques, and Louis Pauwels, *The Morning of the Magicians* (Anthony Gibbs and Phillips, London, 1963; Mayflower Books, London, 1971).

BERGMANN, Peter G., *The Riddle of Gravitation* (John Murray, London, 1969).

BERNAL, J. D., *The World, the Flesh and the Devil* (E. P. Dutton, London, 1929; Cape, London, 1970).

BIRKENHEAD, the Earl of, *The World in 2030* (Hodder and Stoughton, London, 1930).

BONO, Philip, with Kenneth Gatland, *Frontiers of Space* (Blandford Press, London, 1969).

BOVA, Ben, *The New Astronomies* (J. M. Dent, London, 1973).

BUTLER, S. T., and H. Messel (eds), *Apollo and the Universe* (Pergamon Press, Oxford, 1968).

CALDER, Nigel, *Violent Universe* (B.B.C. Publications, London, 1969).

CAMERON, A. G. W. (ed.), *Interstellar Communication* (W. A. Benjamin, New York, 1963).

CHEMICAL RUBBER COMPANY of Cleveland, Ohio, *Handbook of Tables for Mathematics* (4th edition, 1970), edited by Samuel L. Selby.

CLARK, Kenneth, *Civilisation* (B.B.C. Publications and John Murray, London, 1969 and 1971).

CLARKE, Arthur C. (ed.), *The Coming of the Space Age* (Gollancz, London, 1967; Panther Books, London, 1970).

—— *The Promise of Space* (Hodder and Stoughton, 1968; Pelican Books, Harmondsworth, 1970).

—— *Man and Space* (Life Science Library, New York, 1964, revised edition 1970).

—— *Report on Planet Three and Other Speculations* (Gollancz, London, 1972).

CLARKE, Robin, *The Science of War and Peace* (Jonathan Cape, London, 1971).

COLE, Dandridge M., and Donald W. Cox, *Islands In Space* (Chilton Books, New York, 1965).

CROMBIE, A. C., *Augustine to Galileo* (Harvard University Press, 1953).

DALE, Ella, and Donald Michie (eds), *Machine Intelligence*, 2 vols (Oliver and Boyd, London, 1968).

DE CHARDIN, Teilhard, *The Phenomenon of Man* (Collins, London, 1959).

—— *The Future of Man* (Collins, London, 1964).

DESSEL, Norman F., Richard B. Nehrich and Glenn I. Voran, *Atomic Light Lasers* (Oak Tree Press, London, 1969).

DEWITT, Cecile M., and John A. Wheeler (eds), *Battelle Rencontres: 1967 Lectures in Mathematics and Physics* (W. A. Benjamin, New York, 1968).

DOLE, Stephen H., *Habitable Planets for Man* (Elsevier Publishing Co., New York, 1970).

EHRENSVARD, Gosta, *Man on Another World* (Chicago University Press, 1965).

EINSTEIN, Albert, *The Meaning of Relativity* (Methuen, London, 1922).

ELLIS, G. F. R., with S. W. Hawking, *The Large Scale Structure Space-Time* (Cambridge University Press, 1973).

ENCYCLOPEDIA BRITANNICA, *1972 Year Book of Science and The Future*.

FARRINGTON, Benjamin, *Francis Bacon: Philosopher of Industrial Science* (Macmillan, London, 1973; Haskell House, New York, 1973).

FERDMAN, Saul (ed.), *The Second Fifteen Years in Space* (American Astronautical Society, 1973).

GAMOW, George, *A Star Called the Sun* (Penguin, Harmondsworth, 1964).

—— *The Creation of the Universe* (Viking Press, New York, 1952; New American Library, 1957; Bantam Pathfinder, New York, 1965).

GARDNER, Martin, *Relativity for the Million* (Macmillan, New York, 1962; Pocket Books, New York, 1962).

—— *The Ambidextrous Universe* (Allen Lane, London, 1967; Pelican Books, Harmondsworth, 1970).

GATLAND, Kenneth, *Robot Explorers* (Blandford Press, London, 1972).

—— *The Frontiers of Knowledge* (Allan Wingate, London, 1974).

GLASSTONE, Samuel, *Sourcebook on Atomic Energy* (Van Nostrand, Princeton, 1967).

GOOD, I. J. (ed.), *The Scientist Speculates* (Heinemann, London, 1962).

HALDANE, J. B. S., *Daedalus or Science and the Future* (Kegan Paul, Trench, Trubner, London, 1924).

HENRY, George O., *Tomorrow's Moon* (Little, Brown, Boston, 1969).

HODGE, Paul W., *Concepts of the Universe* (McGraw-Hill, New York, 1969).

HOLT, Michael, and D. T. E. Marjoram, *Mathematics in a Changing World* (Heinemann, London, 1973).

HOYLE, Sir Fred, *The Nature of the Universe* (Basil Blackwell, Oxford, 1960).

—— *Of Men and Galaxies* (Washington University Press, 1966).

—— *Galaxies, Nuclei and Quasars* (Heinemann, London, 1966).

—— *The New Face of Science* (New American Library, in association with World Publishing Co., New York, 1971).

JAFFE, Bernard, *Michelson and the Speed of Light* (Heinemann, London, 1960).

JASTROW, Robert, *Stars, Planets and Life* (Heinemann, London, 1968).

—— and S. I. Rascol (eds), *The Venus Atmosphere* (Gordon and Breach, London, 1969).

JEANS, Sir James, *The Universe Around Us* (Cambridge University Press, 1930).

JOHN, Laurie (ed.), *Cosmology Now* (B.B.C. Publications, 1973).

KAHN, F. D., and H. P. Palmer, *Quasars* (Manchester University Press, 1967).

KAHN, Herman, *On Thermonuclear War* (Princeton University Press, 1961).

—— and Anthony J. Wiener, *The Year 2000* (Collier-Macmillan, London, 1969).

KAPLAN, S. A., *Extraterrestrial Civilisations* (Izdatel'stvo 'Nauka' Glavnaya Redaktsia Fiziko-Matematicheskoi Literatury, Moscow, 1969; Israel Programme for Scientific Translations, Jerusalem, 1971).

KAUFMANN, William J., *Relativity and Cosmology* (Harper and Row, New York, 1973).

KLASS, Philip J., *Secret Sentries in Space* (Random House, New York, 1971).

KLAUDER, John R. (ed.), *Magic Without Magic: John Archibald Wheeler* (W. H. Freeman, San Francisco and Reading, 1973).

KLEIMAN, Louis A., *Project Icarus* (Massachusetts Institute of Technology Press, 1968).

KUHN, Thomas S., *The Structure of Scientific Revolutions* (University of Chicago Press, 1962 and 1965).

LEACH, Gerald, *The Biocrats* (Jonathan Cape, London, 1970).

LE GROS CLARK, Sir Wilfred, *History of the Primates* (Trustees of the British Museum, Natural History, 1970).

LEONARD, Jonathan N., and Carl Sagan, *Planets* (Life Science Library, 1966, 1967 and 1970).

LESSING, Lawrence, *DNA: At the Core of Life Itself* (Collier-Macmillan, London, 1967).

LEVINSON, Alfred A., and S. Ross Taylor, *Moon Rocks and Minerals* (Pergamon Press, Oxford, 1971).

LEWIS, Richard S., and Eugene Rabinowitch (eds), *Men in Space* (Medical and Technical Publishing, London, 1970).

LEY, Willy, *Watchers of the Skies* (Sidgwick and Jackson, London, 1964).

—— *Beyond the Solar System*. Paintings by Chesley Bonestell (Viking Press, New York, 1964).

LINDAMAN, Edward B., *Space: A New Direction for Mankind* (Harper and Row, New York, 1969).

LUNAN, Duncan, *Man and the Stars* (Souvenir Press, London, 1974).

MADDOX, John, *The Doomsday Syndrome* (Macmillan, London, 1972).

MALINA, Frank J. (ed.), *Applied Sciences Research and Utilisation of Lunar Resources* (Pergamon Press, Oxford, 1970).

MALTHUS, Thomas, *Essay on the Principle of Population As it Affects the Future Improvement of Society With Remarks on the Speculations of Mr. Godwin, M. Condorcet, and Other Writers* (London, 1798; second edition, much revised, 1803).

MARSHACK, Alexander, *The Roots of Civilisation* (Weidenfeld and Nicolson, London, 1972).

MARSHAK, R. E. (ed.), *Perspectives in Modern Physics* (Interscience Publishers, New York, 1966).

MASON, Brian, and William G. Melson, *The Lunar Rocks* (Wiley-Interscience, London, 1970).

MEADOWS, Dennis L., Donella H. Meadows, Jorgen Randers, and William W. Behrens and the Club of Rome, *The Limits to Growth* (Universe Books, New York, 1972).

MICHELMORE, Peter, *Einstein: Profile of the Man* (Dodd, Mead, New York, 1962).

MISNER, Charles W., with Kip S. Thorne and John A. Wheeler, *Gravitation* (W. H. Freeman, San Francisco and Reading, 1973).

MOORE, Patrick, *Atlas of the Universe* (Mitchell Beazley, London, 1970).

—— *Guide to the Planets* (Lutterworth Press, London, 1971).

—— *Challenge of the Stars*, Paintings by David A. Hardy (Mitchell Beazley with Sidgwick and Jackson, London, 1972).

MORGENSTERN, Oskar, and John von Neumann, *Theory of Games and Economic Behavior* (Princeton University Press, 1953).

MOSZKOWSKI, Alexander, *Conversations with Einstein* (Sidgwick and Jackson, London, 1973).

NICOLSON, Iain, *Exploring the Planets* (Paul Hamlyn, London, 1970 and 1972).

ORDWAY, Frederick A. (ed.), *Advances in Space Science and Technology*, Vol. 10 (Academic Press, London and New York, 1970).

PARNOV, E. I., *At the Crossroads of Infinities* (Mir, Moscow, 1971).

PEARCE WILLIAMS, L. (ed.), *Relativity Theory: Its Impact on Modern Thought* (John Wiley, New York, 1968).

PEEK, Bertrand M., *The Planet Jupiter* (Faber and Faber, London, 1959).

PFEIFFER, John, *The Cell* (Life Science Library, 1965).

PLATO, *The Republic*. Translated by Paul Shoey, 2 Vols (Heinemann, London, 1953).

PYKE, Magnus, *The Science Century* (John Murray, London, 1967).

ROHR, Hans, *The Radiant Universe* (Frederick Warne, London, 1972).

RORVIK, David M., *As Man Becomes Machine* (Doubleday, New York, 1971).

ROSENBERG, Jerry M., *The Computer Prophets* (Collier-Macmillan, London, 1969).

ROTH, Gunter D., *The System of Minor Planets* (Faber and Faber, London, 1962).

RUSSELL, Bertrand, *A History of Western Philosophy* (Allen and Unwin, London, 1946).

—— *The ABC of Relativity* (Allen and Unwin, London, 1958).

RUZIC, Neil P., *The Case for Going to the Moon* (Putnam, New York, 1965).

—— *Where the Winds Sleep* (Doubleday, New York, 1970).

SAGAN, Carl, and Iosif Schlovskii, *Intelligent Life In the Universe* (Holden-Day, San Francisco, 1966).

SAGAN, Carl, *The Cosmic Connection* (Doubleday, New York, 1973; Hodder and Stoughton, London, 1974).

SCHURR, Sam H. (ed.), *Energy, Economic Growth and the Environment* (Published for Resources for the Future, Inc., by the Johns Hopkins University Press, Baltimore and London, 1972).

'SCIENTIFIC AMERICAN' Reprints: *Lasers and Light* (edited by Arthur L. Schlawlow, 1969), and *Frontiers in Astronomy* (edited by Owen Gingerich, 1970).

SNEATH, P. H. A., *Planets and Life* (Thames and Hudson, London, 1970).

STEINHOFF, Ernst A. (ed.), *Organising Space Activities for World Needs* (Pergamon Press, Oxford, 1971).

STRONG, James, *Flight to the Stars* (Temple Press Books, London, 1965).

SOULE, George, *The Shape of Tomorrow* (New American Library, 1958).

SULLIVAN, Walter, *We Are Not Alone* (McGraw-Hill, New York, 1966).

TAYLOR, John, *Black Holes: The End of the Universe* (Souvenir Press, London, 1973).

TERLETSKII, Yakov P., *Paradoxes in the Theory of Relativity* (Plenum Press, New York, 1968).

TOFFLER, Alvin, *Future Shock* (The Bodley Head, London, 1970).

TOVMASYAN, G. M. (ed.), *Extraterrestrial Civilisations* (Izdatel'-stvo Akademii Nauk Armyanskoi SSR, 1965; Israel Programme for Scientific Translations, Jerusalem, 1967).

TSIOLKOVSKY, Konstantin E., *Selected Works* (Mir, Moscow, 1968).

VON NEUMANN, John, *The Computer and the Brain* (Yale University Press, 1958).

WATSON, James D., *The Double Helix* (Weidenfeld and Nicolson, London, 1968; Penguin Books, Harmondsworth, 1968).

WHEELER, John A. (ed.), *Geometrodynamics* (Academic Press, London and New York, 1962).

WHITNEY, Charles A., *The Discovery of our Galaxy* (Alfred A. Knopf, New York, 1971).

WHITROW, G. J., *What is Time?* (Thames and Hudson, London, 1972).

Notes and References

CHAPTER 1: THE NEW ATLANTIS

1. Quoted in Bertrand Russell's *A History of Western Philosophy* (1946), p. 458.
2. Ibid., p. 46.
3. Lord Macaulay, essay on Bacon. From his *Critical and Historical Essays*, edited by Lady Trevelyan (Longmans, Green, London, 1866).
4. Seneca, *Epistulae Morales*, part 90.
5. Ibid.
6. *A History of Western Philosophy*, p. 153.
7. According to Plutarch, Archimedes 'did not regard the inventing of his machines as an object worthy of his serious studies, but only reckoned them among the amusements of geometry'. See Plutarch's *Life of Marcellus*. (Marcellus was the Roman general against whom Archimedes defended Syracuse from 215 to 212 B.C.)
8. Plato's *Republic*, book 7. These opinions are put into the mouth of 'Socrates', but it is generally considered that Plato was attributing his own ideas to his dead hero. See *A History of Western Philosophy*, p. 152.
9. Bacon, *On the Advancement of Learning*, book 3, chapter 4.
10. See Plato's *Phaedrus*. Plato put these opinions into the mouth of an ancient king of Egypt, but again it is evident from the context that they were his own. This, at least, was the view of Quintillian. See his *Institutio Oratio*, book 11.
11. *On the Advancement of Learning*, book 5, chapter 5.
12. Plato's *Republic*, book 3.
13. *On the Advancement of Learning*, book 4, chapter 2.
14. Dante suggested that hypocrites would be punished by being planted upside down in boiling oil for eternity. *The Inferno*, part 9.
15. William H. Prescott, *The Conquest of Peru*, vol. 3, chapter 7.
16. Quoted in *A History of Western Philosophy*, p. 550.
17. See, for example, J. C. Stobart, *The Grandeur that was Rome* (1912), p. 29.
18. Macaulay's essay probably gives the best account of Bacon's short-comings as a politician. Lytton Strachey, in his *Elizabeth and Essex* (1928), takes a most severe view, portraying Bacon as a monster of in-gratitude and treachery.
19. The opinion of Russell, *A History of Western Philosophy*, p. 564.
20. *The New Atlantis*, edited by Harold Osborne (1937). Also, vol. 3 of *The Works of Francis Bacon*, edited by J. Spedding, R. L. Ellis, and D. D. Heath (1857). I. F. Stone gives an interesting review and summary of the novel in 'Bacon's New Atlantis: Blueprint for Progress' (*Futures*, vol. 4, no. 3, September 1972).

21. Abraham Cowley, ode *To The Royal Society*.
22. Quoted by I. F. Stone in 'Bacon's New Atlantis'.
23. Macaulay's essay on Bacon.
24. N. S. Kardashev, 'Transmission of Information by Extraterrestrial Civilisation', *Soviet Astronomy*, vol. 8, p. 217, 1964. See also Kardashev's contribution to *Extraterrestrial Civilisation*, edited by S. A. Kaplan 1969), pp. 22–8.
25. Sir James Jeans, *The Universe Around Us* (1930), p. 343.

CHAPTER 2: NO LIMITS TO GROWTH

1. *The New York Review of Books*, Nov. 30, 1972.
2. See, for example, Fred Hoyle, *The New Face of Science* (1971, pp. 97–112.
3. For detailed arguments against the Club of Rome's report, see the following editorials in *Nature*: 'Another Whiff of Doomsday', vol. 238, March 10, 1972: 'Homilies for the Club of Rome', vol. 238, August 4, 1972; 'More Coals of Fire for the Club of Rome', vol. 239, Sept. 29, 1972; 'Almost the last Word on the Club of Rome', vol. 242, March 16, 1973. Also W. Beckerman, 'The Myth of Environmental Catastrophe', *National Review*, Nov. 24, 1972.
4. T. W. Oerlemans, M. M. J. Tellings and H. de Vries, 'World Dynamics: Social Feedback may give Hope for the Future', *Nature*, vol. 238, August 4, 1972.
5. *The Limits to Growth*, Model No. 42, p. 140. This extraordinary upward curve was noticed by Carl Kaysen. See Kaysen's article, 'The Computer that Printed out WOLF', *Foreign Affairs*, July 1972.
6. Quoted by Wilfred Beckerman, a fellow-member of the Royal Commission on Environmental Pollution, in 'The Myth of Environmental Catastrophe'.
7. Report on '*The Limits to Growth*'. A study by a special Task Force of the World Bank, Sept. 1972.
8. These figures are from Stephen Dole's *Habitable Planets for Man*, pp, 84–5. Other specialists have produced similar figures. See, for example. George Gamow, *A Star called the Sun* (1964), pp. 138–51; S. S. Huang, 'Life Outside the Solar System', *Scientific American*, April 1960; Patrick Moore, *Atlas of the Universe* (1970), p. 197.
9. See all the sources listed in Note 8. Estimates of the Sun's future on the main sequence have varied from 5 billion to 8 billion years. One 'pessimist' believes that elemental impurities in the Sun's core will interfere with its balance of energy, and bring about a temporary doubling of the world mean temperature of 58 degrees F. to about 120 degrees within 1 billion years. Such an increase in heat could make life very difficult on Earth. See E. J. Opik, 'Climate and the Changing Sun', *Scientific American*, June 1958.
10. Chet Holifeld, *Hearings before the Special Subcommittee on Military Operations, Committee on Government Operations, Civil Defense* (1960 and 1961). *Hearings on Biological and Environmental Effects of Nuclear War* (1959). Holifeld's work is quoted in Robin Clarke's gloomy book, *The Science of War and Peace* (1971), pp. 24–5.
11. From Paul R. Ehrlich's *The Population Bomb* (Pan Books, in association with Friends of the Earth, 1968 and 1971).

12. William and Paul Paddock, *Famine—1975!* (Little, Brown, Boston, 1967).
13. J. B. Calhoun, 'Social Aspects of Population Dynamics', *Journal of Mammalogy*, vol. 33, 1952, pp. 139–59. Calhoun's work is quoted by Robert Ardrey in his *The Social Contract* (1970), p. 184.
14. These figures are from John Maddox's *The Doomsday Syndrome* (1972), p. 53. See also *United Nations Demographic Yearbook, 1969* (United Nations, 1970); and *World Population Prospects, 1965–2000, as Assessed in 1968*, Working Paper No. 37, Population Division, U.N Department of Economic and Social Affairs (United Nations, 1970).

CHAPTER 3: THREE ASSUMPTIONS

1. Arthur C. Clarke's epilogue to *First on the Moon*, by Gene Farmer and Dora Jane Hamblin (Michael Joseph, London, 1970).
2. These anecdotes are an expansion of a list which I compiled for an article in the *Daily Telegraph Magazine*, April 30, 1971. I have added some from the article 'Naysayers Never Die', in *Aviation Week and Space Technology*, March 22, 1971.
3. *The New York Times*, Jan. 13, 1920.
4. Comments by Arthur C. Clarke, a founder-member of the British Interplanetary Society, from his essay, 'Memoirs of an Armchair Astronaut (Retired)'.
5. Sir George Edwards, Presidential Address; *Journal of the Royal Aeronautical Society*, vol. 62, April 1958.
6. Dyson expanded this lecture into the article 'Discussion Paper; Interstellar Transport', *Annals of the New York Academy of Science*, vol. 163, p. 347, 1969.
7. These anecdotes are from *The Twelve Caesars* of Suetonius and Tacitus's *Annals of Imperial Rome*. It should be pointed out that some modern writers, Italians in particular, object to these stories about Nero. One of them asserts that the fire was an accident and that Nero's supposed song was in fact an exhortation to the fire-fighters. Others allege that Suetonius and Tacitus maligned Nero to please the contemporary Antonine emperors, in the same way that Thomas More and Shakespeare are said to have blackened Richard III's character to please the Tudors. Personally, I believe that the testimony of ancient writers is preferable to modern ones in such matters, since they are much nearer to the event in time.
8. Carl Sagan and Iosif Schlovskii, *Intelligent Life in the Universe* (1966), pp. 475–6. They based their calculations on a suggestion that exploding stars could by themselves, blow up their neighbouring stars, thus starting a chain reaction through a galaxy. See G. R. Burbidge, 'Galactic explosions as Sources of Radio Emission', *Nature*, No. 4781, June 17, 1961.
9. See the discussion surrounding the 'Resolution on Scientific Freedom', *Encounter*, Dec. 1972. Also, Arthur R. Jensen, 'How Much Can We Boost IQ and Scholastic Achievement?', *Harvard Educational Review*, Winter 1969; and Jensen's book, *Genetics and Education* (Methuen, London, 1972). For a more popularly written account of the victory of the genetic school of thought over the environmental, see T. Alexander, 'The Social Engineers Retreat Under Fire', *Fortune*, Oct. 1972.

10. I have been unable to identify the author of this prose. I know many people who can remember learning it at school, but I have not been able to find it in any published book on Henry VIII. Possibly it was written by some schoolmaster to teach memorization and never published.
11. Gerald Leach, *The Biocrats* (1970), p. 95.
12. Sir Peter Medawar, 'Science and the Sanctity of Life: An Examination of Current Fallacies', *Encounter*, Dec. 1966.
13. The average annual economic growth-rate of industrial countries during the decade from 1960 to 1970 was 5·3 per cent. See Table 173 in *Social Trends*, published by the Central Statistical Office (H.M. Stationery Office, London, 1972).
14. Shackleton's interview with Dr Anthony Michaelis, *Daily Telegraph*, Dec. 20, 1972.

CHAPTER 4: THE BECKONING MOON

1. L. Caretti (ed.), *Orlando Furioso* (1954).
2. Alexander Pope, *The Rape of the Lock*. See selections of his poetry in Everyman's Library, edited by Bonamy Dobrée (1956).
3. Alexander Marshack, *The Roots of Civilisation* (1972), chapters 1–5.
4. Ibid. p. 140.
5. See, for example, accounts of the first and second Lunar Science Conferences, from 1969 to 1972; published by the Massachusetts Institute of Technology in several volumes. Also, Patrick Moore's *Guide to the Planets* (1971), and Carl Sagan and Jonathan Norton Leonard, *Planets* (1970).
6. *Time* magazine, October 25, 1971, p. 59. *Daily Telegraph*, October 16, 1971.
7. NASA Technical Memorandum No. TMX-58061, presented to the American Institute of Chemical Engineers National Symposium at Houston, Texas, March 1–6, 1971.
8. Brian Mason and William G. Melson, *The Lunar Rocks* (1970), p. 101.
9. This idea resulted from author's joint interview with Moore and Michaelis.
10. This article is reprinted in Arthur C. Clarke's *The Promise of Space* (1968 and 1970).
11. F. Zwicky, 'Systems for Extracting Elements and Chemical Compounds from Lunar Materials needing Manned Operations on the Moon'. Contribution to *Applied Sciences Research and Utilisation of Lunar Resources*, edited by F. J. Malina (1970).

CHAPTER 5: THE LUNARIANS

1. *Time* magazine, April 12, 1971, p. 44.
2. *Sunday Telegraph*, May 7, 1972.
3. This list is based on a study by George E. Henry, in *Tomorrow's Moon* (1969).
4. A calculation from Hans Rohr's *The Radiant Universe* (1972), p. 31.
5. A prediction from Neil Ruzic's *Where the Winds Sleep*, 1970, p. 82.
6. Plate 75 of Rohr's *The Radiant Universe*.
7. R. Barkan, 'The Laser Goes into Battle', *New Scientist*, July 13, 1972.
8. See, for example, P. M. Winslow and D. V. McIntyre, 'Adhesion of Metals in the Space Environment', *Journal of Vacuum Science and*

Technology, vol. 3, no. 2, 1966; and P. J. Silverman and G. P. Newton, 'Relation of Measured Outgassing Pressures to Surface Adsorption in Satellite Borne Pressure Gauges', *Journal of Vacuum Science and Technology*, vol. 7, no. 2, 1970.

9. Neil Ruzic, *The Case for Going to the Moon*, 1965. See also the Encyclopedia Britannica's *Yearbook of Science and the Future, 1972*, pp. 16–17.

10. H. F. Wuenscher, 'Manufacting in Space', *New Scientist*, September 10, 1970. Also K. A. Ehricke, 'Extraterrestrial Imperative', *Bulletin of the Atomic Scientists*, November 1971.

11. Ibid.

12. Arthur C. Clarke, *Report on Planet Three and Other Speculations* (1972), p. 236.

13. Ibid, p. 149.

14. See, e.g., Ehricke, 'Extraterrestrial Imperative'. Also, S. von Hoerner, 'Population Explosion and Interstellar Expansion', contribution to Sussman and Scheibe (eds.), *Einheit und Vielheit* (Vandenhoeck and Ruprecht, Gottingen, 1972).

15. See a discussion by Clarke in *The Promise of Space*, pp. 61–89.

16. Clarke, 'Electromagnetic Launching as a Major Contribution to Space Flight', *Journal of the British Interplanetary Society*, November 1950.

17. The term 'lunartron' was invented in 1962 by William J. D. Escher of the Marshall Space Flight Center, Huntsville, Alabama.

18. See Kenneth Gatland and Philip Bono, *Frontiers of Space* (1969). See also a pamphlet by William Escher, *The Synerjet Engine*, published in November 1971 by Escher Technology Associates, P.O. Box 189, St Johns, Michigan 48879.

19. These underground Lunar dwellings are based on a study by Mitchell R. Sharpe. See Britannica's *Yearbook of Science and the Future, 1972*, p. 13.

20. Interview with von Braun on American television, January 6, 1972.

21. *The Promise of Space*, p. 213.

CHAPTER 6: VENUS, THE HELL-WORLD .

1. Our knowledge of Venus's atmosphere comes not only from radar measurements and observations by optical and radio telescopes on Earth, but also from the unmanned American Mariner and Russian Venera spacecraft that have orbited the planet, and on some occasions landed on it. For a popularly-written account of these, see Patrick Moore's *Guide to the Planets* (1971). For more technical and more detailed summaries, see R. Jastrow and S. I. Rasool (eds), *The Venus Atmosphere—Papers from the Second Arizona Conference on Planetary Atmosphere* (1969). For other general discussions, see Carl Sagan and Jonathan Norton Leonard, *Planets* (1966, 1967 and 1970). The following articles are also useful: J. Nikander, 'Some Problems Posed by the Planet Venus', *Spaceflight*, October 1970; R. N. Watts, 'The Exploration of Venus', *Sky and Telescope*, February 1971; V. S. Avduevsky, M. Ya. Marov and M. K. Rozhdestvensky, 'A Tentative Model of the Venus Atmosphere based on the Measurements of Veneras 5 and 6', *Journal of the Atmospheric Sciences*, vol. 27, July 1970; D. A. de Wolf, 'Atmospheric Turbulence on Venus; Venera 4, 5, 6 and Mariner 5 Estimates', *Journal of Geophysical Research*, vol. 76, May 1, 1971.

2. Sagan and Leonard, *Planets*, page 111.
3. The effects of this super-refractivity are described with vivid diagrams by Von R. Eshleman in 'The Atmospheres of Mars and Venus', *Scientific American*, March 1969.
4. As two astronomers have stated, 'We believe that the chief differences between Earth and Venus can be explained by the single circumstance that Venus was formed 30 per cent closer to the Sun.' (S. I. Rasool and C. de Bergh, 'The Runaway Greenhouse and the Accumulation of CO_2 in the Venus Atmosphere', *Nature*, vol. 226, June 13, 1970.)
5. A thorough discussion of this problem is given by Su-Shu Huang, in 'The Sizes of Habitable Planets', his contribution to *Interstellar Communication*, edited by A. G. W. Cameron (1963).
6. J. B. Pollack, 'A Nongrey Calculation of the Runaway Greenhouse: Implications for Venus's Past and Present'. *Icarus*, vol. 14, p. 295, 1971.
7. Carl Sagan and Iosif Schlovskii, *Intelligent Life in the Universe* (1969), Sagan's chapter on Venus in this book is reprinted in part in *Farewell, Fantastic Venus!—A History of the Planet Venus in Fact and Fiction*, edited by Brian W. Aldiss and Harry Harrison (1968 and 1971).
8. C. Sagan, 'The Planet Venus', *Science*, vol. 133, March 24, 1961.

CHAPTER 7: MAKING IT RAIN IN HELL

1. Dr P. H. A. Sneath, a biologist at Leicester University, describes the astonishing hardiness of algae in chapter 2 of his *Planets and Life* (1970).
2. 'Watermelon snow' is one of the strangest phenomena which a mountain traveller can see. It is snow which has been tinged pink by algae. A photograph of this pink snow appears on p. 105 of *The Mountains* (Life Nature Library, 1962 and 1967), by Lorus J. Milne and Margery Milne,
3. Sneath, *Planets and Life*, pp. 78–87. Sagan and Leonard, *Planets*, pp. 38–41.
4. 'The arrival of micro-organisms on Venus may have a dramatic effect on the planet.' Sagan and Leonard, *Planets*, p. 117.
5. For full descriptions of these experiments and their results, see J. Seckbach and F. W. Libby, 'Vegetative Life on Venus? Or Investigations with Algae which Grow under Pure Carbon Dioxide in Hot Acid Media at Elevated Pressures', *Space Life Sciences*, vol. 2, 1970, pp. 121–43; and J. Seckbach, F. A. Baker and P. M. Shugarman, Algae Thrive under Pure Carbon Dioxide', *Nature*, vol. 227, August 15, 1970.
6. Poul Anderson's short story, 'The Big Rain', appeared in the magazine *Astounding Science Fiction* in 1955. This story is reprinted by Aldiss and Harrison in *Farewell, Fantastic Venus!*
7. Sagan and Leonard, *Planets*, p. 117.
8. One such scheme is described in Arthur C. Clarke's novel *The Sands of Mars* (1951 and 1954). In this story, the Martian moon Phobos was fitted out with a meson reactor, emitting radiation on plants already growing on Mars's surface, accelerating their growth, and thereby increasing the proportion of oxygen in that planet's atmosphere. Dr Fred S. Singer, an American physicist, has suggested that the Earth's Moon might be induced to hold an atmosphere if its density, and consequently its gravitational attraction, could be increased. This would be achieved by exploding a nuclear device at the Moon's centre, causing the Moon to implode (*Daily Telegraph*, July 21, 1969). But this plan might face

strenuous opposition on aesthetic grounds. It would also destroy all the Moon's industries.

9. N. Fukuta, T. L. Wang and W. F. Libby, 'Ice Nucleation in a Venus Atmosphere', *Journal of the Atmospheric Sciences*, vol. 26, September 1969. Their belief in the existence of ice crystals in the Venus atmosphere provoked a dissent by J. S. Lewis in the same journal, vol. 27, March 1970. Fukuta, Wang and Libby replied vigorously to Lewis in the same issue. They seem to represent the majority view; see also, M. Bottema; W. Plummer, J. Strong and R. Zander, 'The Composition of the Venus Clouds and Implications for Model Atmospheres', *Journal of Geophysical Research*, vol. 70, September 1, 1965; and C. Sagan and J. B. Pollack, 'Anisotropic Nonconservative Scattering and the Clouds of Venus', ibid., vol. 72, January 15, 1967. Both of these latter articles strongly support the theory that ice crystals are present.

10. H . Morowitz and C. Sagan, 'Life in the Clouds of Venus?', *Nature*, vol. 215, September 16, 1967.

11. Sagan and Leonard at first imagined that creatures of this kind might inhabit the murky atmosphere of Jupiter. An artist's conception of them, pictures with huge bulbous eyes, appears on p. 190 of *Planets*.

12. Olaf Stapledon, *Last and First Men* (Methuen 1930, Penguin, 1968). I have given a somewhat abbreviated excerpt.

13. Sagan himself, in his original 1961 article, actually makes this warning against any hasty implementation of his own plan. It should be done, he said, 'only after the present Cytherean (i.e. Venus) environment has been thoroughly explored, to prevent the irreparable loss of unique scientific information'. Some people have commented that the only way to restrain would-be Venus colonists would have been to suppress the plan in the first place. But Sagan, as a true scientist, was unable to do this. See my own comments in 'After the Rain', *Daily Telegraph Magazine*, no. 340, April 30, 1971.

14. 'Civilisation means something more than energy and will and creative power. How can I define it? Well, very shortly, a sense of permanence. The wanderers and the invaders were in a continual state of flux. They didn't feel the need to look forward beyond the next march or the next voyage or the next battle. Almost the only stone building that has survived from the centuries after the Mausoleum of Theoderic is the Baptistry at Poitiers. It is pitifully crude. But at least this miserable construction is meant to last. It isn't just a wigwam. Civilised man, or so it seems to me, must feel that he belongs somewhere in space and time; that he consciously looks forward and looks back.' Kenneth Clark, *Civilisation* (1969), pp. 14–17.

15. C. Sagan, 'The Long Winter Model of Martian Biology: A Speculation', *Icarus*, Dec. 1971.

CHAPTER 8: THE ASTRONAUT'S SHRUNKEN HEAD

1. E. Purcell, 'Radio Astronomy and Communication through Space', contribution to *Interstellar Communication*, A. G. W. Cameron, ed. (1963).

2. Alexander Pope wrote the Newton couplet in the 18th century. J. C. Squire answered Pope with the second.

3. Bernard Jaffe, *Michelson and the Speed of Light* (1961), p. 95.

4. The first limerick was written by E. R. Reginald Buller, a botanist at the University of Manitoba. The sequel is the work of J. H. Fremlin, a physicist at Birmingham University. See Fremlin, 'Newton, Relativity and Free Will', *University of Birmingham Review*, Autumn 1966.
5. These events are described vividly in Peter Michelmore's biography, *Einstein, Profile of the Man* (1962).
6. 'A Mystic Universe'. Editorial in the *New York Times*, January 28, 1928.
7. R. T. Powers and J. A. Wheeler, 'Mendeleev and the Chemical Orbit', paper presented on September 16, 1969, to the Accademia delle Scienze di Torino and the Accademia Nazionale dei Lincei, to commemorate the 100th anniversary of Dimitri Mendeleev's publication of the periodic law of elements.

CHAPTER 9: VOYAGE THROUGH A HIDDEN UNIVERSE

1. Isaac Asimov, *The Stars Like Dust* (Panther Books, London, 1958).
2. R. W. Fuller and J. A. Wheeler, 'Causality and Multiply-Connected Space-Time', *Physical Review*, vol. 128, October 15, 1962.
3. As Hoyle says in his popular book, *The Nature of the Universe* (1960), p. 89, 'Matter that already exists causes new matter to appear. This may seem a very strange idea, and I agree that it is, but in science it does not matter how strange an idea may seem so long as it works—that is to say, so long as the idea can be expressed in a precise form, and so long as its consequences can be found in agreement with observation.' Hoyle has often expressed this idea in mathematical form. See Hoyle and S. V. Narlikar, 'On the Formation of Elliptical Galaxies', *Proceedings of the Royal Society A*, vol. 290, 1966.
4. Gamow describes this terrific event almost like an eye-witness in his lively book, *The Creation of the Universe* (1952, 1957 and 1965).
5. J. A. Wheeler, 'Our Universe: the Known and the Unknown', *American Scientist*, Spring 1968.
6. B. E. Turnrose and H. J. Rood, 'On the Hypothesis that the Coma Cluster is Stabilised by a Massive, Ionized Intergalactic Gas', *Astrophysical Journal*, vol. 159, March 1970.
7. 'Year of the Black Hole'; Wheeler interviewed by Jeremy Campbell in the London *Evening Standard*, Jan. 10, 1973. Also, 'The Black Hole of the Universe'; Wheeler interviewed by Lawrence B. Chase, *Intellectual Digest*, Dec. 1972. This latter article was reproduced from *University*, a *Princeton Quarterly*, Summer 1972.
8. J. Weber, 'Gravitational Radiation Experiments', *Physical Review Letters*, vol. 24, Feb. 9, 1970.
9. J. Weber, 'Anisotropy and Polarisation in the Gravitational-Radiation Experiments', *Physical Review Letters*, vol. 25, July 20, 1970. An excellent popular account of Weber's experiments is given by Tom Alexander, 'Mystery of the Gravity Waves', in the *Nature/Science Annual* of 1971, p. 117 (Time-Life Books, 1971). At a conference on relativity at All Souls College, Oxford, on April 27, 1973, Weber's results were criticized by some experimenters who had failed to duplicate them. Some scientists accused him of approaching his experiments with an 'unconscious human bias'. See an account of this conference in *Nature*, vol. 243, May 11, 1973. Weber was unmoved by this attack. He said in a telephone interview, 'I don't make decisions about what is good data

or what is bad. The computer points out what the coincidences are. I have studied a million feet of chart records and I am not moved by the charge of unconscious bias.' Weber has always been at pains to describe the meticulous care which he took with his experiments. See, for example, his letter to *Nature* (vol. 240, Nov. 3, 1972), 'Computer Analyses of Gravitational Radiation Detector Coincidences'. D. Lynden-Bell, who defended Weber at the All Souls conference, said that *he* would only worry about the absence of gravitational waves when instruments a trillion times more sensitive than Weber's had failed to detect them.

10. G. B. Field, M. J. Rees and D. W. Sciama, 'The Astronomical Significance of Mass Loss by Gravitational Radiation', *Comments on Astrophysics and Space Physics*, vol. 1, p. 187, 1969. See also D. W. Sciama, 'Is the Galaxy Losing Mass on a Time Scale of a Billion Years?', *Nature*, vol. 224, December 27, 1969, and, R. Penrose, 'Black Holes and Gravitational Theory', ibid., vol. 236, April 21, 1972. And R. Penrose 'Black Holes', *Scientific American*, May 1972.

11. R. M. Hjellming, 'Black and White Holes', *Nature Physical Science*, vol. 231, May 3, 1971.

12. The vivid description 'anti-collapse' appears in I. D. Novikov's paper, 'The Replacement of Relativistic Gravitational Attraction by Expansion, and the Physical Singularities during Contraction', *Soviet Astronomy*, March–April 1967.

13. John A. Wheeler, *Geometrodynamics* (1962).

14. J. A. Wheeler and S. Tilson, 'The Dynamics of Space-Time', *International Science and Technology*, Dec. 1963.

15. 'The Dynamics of Space-Time.'

16. 'Our Universe: The known and the Unknown.'

17. 'Causality and Multiply-Connected Space-Time.'

18. K. S. Thorne, 'Gravitational Collapse', *Scientific American*, Nov. 1967.

19. Margaret Burbidge, 'Nuclei of Galaxies and Quasars—White Holes or Black Holes?' Address at the Royal Institution, London, Oct. 20, 1972. Ya B. Zel'dovich and I. D. Novikov, 'Relativistic Astrophysics', *Soviet Physics: Uspekhi*, vol. 7, pp. 763–88, May–June 1965 and ibid., vol. 8, pp. 522–77, Jan–Feb. 1966. Bardeen's work is quoted in the article cited in note 18.

20. Hawking and Ellis, *The Large Scale Structure of Space-Time*, p. 360.

21. Author's interview with Ne'eman, *Daily Telegraph*, May 4, 1970.

22. T. Gold, 'Multiple Universes,' *Nature*, Vol. 242, March 2, 1973.

23. A. Einstein and N. Rosen, 'The Particle Problem in the General Theory of Relativity.' *Physical Review*, vol. 48, July 1, 1935.

24. Alexander Mozkowski, *Conversations with Einstein*, p. 129.

25. See my own account of this experiment; 'Man Who Was Younger than his own Children,' *Daily Telegraph*, Oct. 29, 1972.

26. More than 400 articles, books and reports have been published on the feasibility of propelling spacecraft to relativistic speeds. A complete list up to January, 1971, is given by Eugene F. Mallove and Robert L. Forward in their *Bibliography of Interstellar Travel and Communication*, published in the *Journal of the British Interplanetary Society*, vols 27–28, 1974–75. James Strong's 1965 *Flight to the Stars* is also of great value.

CHAPTER 10; THE SEARCH FOR HABITABLE WORLDS

1. Plutarch, *On the Tranquillity of the Mind*.
2. Isaac Asimov in his *Biographical Encyclopedia of Science and Technology*.
3. Stephen Dole, *Habitable Planets for Man*, p. 87.
4. I am indebted for many of these calculations to Dole, pp. 130–33.
5. David Bergamini, *The Universe* (1969), p. 112.
6. *Daily Telegraph*, April 24, 1969.
7. Author's interview with Drake, *Daily Telegraph*, October 6, 1969.
8. James Strong, *Flight to the Stars* (1965), p. 127.
9. Arrian, *The Campaigns of Alexander*, book 7. Also Diodorus, *Bibliotheca Historica*, part 2.

CHAPTER 11: THE VISION OF JOHN VON NEUMANN

1. Hughes Aircraft Company Press Release, March 16, 1972.
2. This conversation appears in David Rorvik's *As Man Becomes Machine: The Evolution of the Cyborg* (1971).
3. *The Prospects for the United Kingdom Computer Industry in the 1970s*. Fourth Report from the Select Committee on Science and Technology, vol. 1. House of Commons Command Paper 621, November 18, 1971.
4. S. Ulam, 'John von Neumann', *Bulletin of the American Mathematical Society*, vol. 64, no. 654, May 1958. The entire May 1958 issue of this journal is devoted to discussions by many authors of von Neumann's life's work, edited by B. J. Pettis and G. B. Price.
5. John von Neumann and Oskar Morgenstern, *Theory of Games and Economic Behaviour* (1953).
6. This famous lecture is reproduced in pages 288–328 of vol. 5 of von Neumann's *Collected Works* (1963).
7. 'The General and Logical Theory of Automata.'
8. Of the millions of words published about DNA and RNA, two of the clearist and simplest accounts appear in John Pfeiffer's *The Cell* (1969), and in Lawrence Lessing's *DNA: At the Core of Life Itself* (1966).
9. Asimov's novels and collections of short stories about robots are entitled, *I. Robot*: *The Rest of the Robots*; *The Caves of Steel*; and *The Naked Sun* (Dobson Books, and Panther, 1964–8).
10. Freeman J. Dyson, 'The Twenty-First Century'. Vanuxem Lecture delivered at Princeton University, February 26, 1970. Unpublished at the time of writing.
11. Ibid.

CHAPTER 12: FLYING CITY-STATES

1. For an account of Project Ozma, see Frank Drake's contribution to *Interstellar Communication*, edited by A. G. W. Cameron (1963). Also F. Drake, 'How Can We Detect Radio Transmissions from Distant Planetary Systems?', *Sky and Telescope*, vol. 19, p. 140, 1959; G. Cocconi and P. Morrison, 'Searching for Interstellar Communications', *Nature*, vol. 184, p. 844, 1959. A good journalistic description of Project Ozma is given in Walter Sullivan's *We Are Not Alone* (1964 and 1966), pp. 196–207.
2. O. Struve, 'Astronomers in Turmoil', *Physics Today*, vol. 13, pp. 22–3, Sept. 1960.

3. F. J. Dyson, 'Search for Artificial Stellar Sources of Infrared Radiation'; contribution to Cameron's *Interstellar Communication*. This paper appeared originally in *Science*, vol. 131, p. 1667, 1959.
4. F. J. Dyson, 'Intelligent Life in the Universe', lecture given at San Francisco, Sept. 18, 1972, under sponsorship of the Astronomical Society of the Pacific, NASA, and the City College of San Francisco. Sent to me by kind courtesy of the author.
5. See, for example, T. K. Fowler and R. F. Post, 'Progress Towards Fusion Power', *Scientific American*, Dec. 1966; W. C. Gough and B. J. Eastlund, 'The Prospects of Fusion Power', *Scientific American*, Feb. 1971; L. Artsimovich, 'The Road to Controlled Nuclear Fusion', *Nature*, vol. 239, Sept. 1, 1972; F. Knebel, interview with Melvin B. Gottlieb of the Princeton Plasma Physics Laboratory, 'Work in Progress', *Intellectual Digest*, Sept. 1972.
6. Sir Fred Hoyle, *The New Face of Science* (1971), pp. 117–18.
7. Space cities have been widely discussed even in the popular press. See, for example, Brian Aldiss, 'City in the Sky', *The Illustrated London News*, May 16, 1970.
8. McDonnell-Douglas Aircraft Corporation, Study Paper No. 4502, May 3, 1967. Also published in *Advances in the Astronautical Sciences*, vol. 23, 'Commercial Utilisation of Space' (Astronautical Society, Publications Office P.O. Box 746, Tarzana, California 91356).
9. See for example, G. Chedd, 'Colonisation at Lagrangea,' *New Scientist*, Oct. 29, 1974; and O'Neill's own papers, 'The Colonisation of Space', *Physics Today*, Sept. 1974, and, 'A Lagrangean Community?' *Nature*, vol. 250, Aug. 23, 1974.
10. Quoted in *New Scientist* article cited in Note 9.
11. For a discussion of electric power systems in giant satellites and space cities, see, J. E. Bore 'Large Space Station Power Stations,' *The Journal Spacecraft and Rockets*, vol. 6, No. 8, Aug. 1969.
12. K. A. Ehricke, 'The Extraterrestial Imperative,' *Bulletin of the Atomic Scientists*, Nov. 1971.
13. K. E. Tsiolkovsky, 'Investigation of World Spaces by Reactive Vehicles', *Vestnik Vozdukhoplavaniya (Herald of Astronautics)*, vol. 9, pp. 7–8, 1912.

CHAPTER 13: BUILDING THE GIANT SPHERE

1. Reinhold Ebertin, *The Combination of Stellar Influences*, quoted in Louis MacNeice's *Astrology* (Aldus Books, London, 1964). I am not, however, suggesting that astrological forecasts should be read for any purpose other than amusement.
2. There are many useful books and articles on our knowledge of Jupiter to date. See, in particular, B. M. Peek, *The Planet Jupiter*; Sagan and Leonard, *Planets*; Patrick Moore, *Guide to the Planets*. Also, F. O. Rice, 'The Chemistry of Jupiter', *Scientific American*, June 1956; K. L. Franklin, 'Radio Waves from Jupiter', *Scientific American*, July 1964; H. Reeves and Y. Bottinga, 'The Deuterium/Hydrogen Ratio in Jupiter's Atmosphere', *Nature*, vol. 238, Aug. 11, 1972; D. S. Evans and W. B. Hubbard, 'Discrepancies in Measurement of the Jupiter Atmospheric Scale Height', *Nature*, vol. 240, Dec. 18, 1972; G. R. A. Ellis, 'Fire Structure of the Jupiter Radio Bursts', *Nature*, vol. 241, Feb. 9, 1973.

3. A spectacular painting by David A. Hardy of a hypothetical manned observatory on Amaltheia appears in Patrick Moore's *Challenge of the Stars*, p. 28.
4. F. J. Dyson, 'The Search for Extraterrestrial Technology', contribution to *Perspectives in Modern Physics*, edited by R. E. Marshak.
5. Author's interview with J. H. Fremlin.
6. J. Strong, 'Trojan Relays: A Method for Radio Communication Across the Solar System', *Wireless World*, March 1967.
7. Dyson, 'The Search for Extraterrestrial Technology'.
8. Ibid.
9. G. R. Burbidge, 'Galactic Explosions as Sources of Radio Emission', *Nature*, vol. 190, June 17, 1961.
10. Sagan and Schlovskii, *Intelligent Life in the Universe* (1966), pp. 475–6.
11. As Dyson puts it in his contribution to Marshak's book, a 'tamed' galaxy would look quite different from the outside to a 'wild' one, 'Starlight, instead of shining wastefully, would be carefully dammed and controlled. Stars, instead of moving at random, would be grouped and organised.' Since nothing of this kind can be seen in our galaxy, he concludes that no advanced alien Phase 3 civilization has yet 'taken it over'.
    For other discussions on galactic technology, see Sagan and Schlovskii, *Intelligent Life in the Universe*, pp. 467–88. Also Sir Fred Hoyle, *The New Face of Science*, pp. 113–18. Also A. Berry, 'And Man Remade the Firmament', *Daily Telegraph Magazine*, no. 342, May 14, 1971.
12. C. D. J. Generales, 'Quest for Cosmic Survival', *New York State Journal of Medicine*, vol. 70, no. 10, May 15, 1970. See also Generales, 'The Interaction Between Man, Earth and Universe', *Medical Opinion and Review*, June 1968.

CHAPTER 14: THE GOD OF SPINOZA

1. An almost unanimous opinion. See, for example, A. G. W. Cameron's *Interstellar Communication* and P. H. A. Sneath's *Planets and Life*.
2. W. E. Le Gros Clark, *History of the Primates* (1970), pp. 119–21.
3. A detailed account of Einstein's disputes with the Rabbis and the Catholics appears in Michelmore's *Einstein: Profile of the Man*.
4. See, for example, Saint Augustine, 'No salvation exists outside the Church', *De Baptismo*, vol. 4, ch. 17, in which he commends Saint Cyprian's statement that 'He cannot have God for his father who has not the Church for his mother.'
5. Arthur C. Clarke, *Report on Planet Three and Other Speculations* (1972).
6. F. J. Dyson, 'Energy in the Universe', *Scientific American*, September 1971.
7. This hypothetical disaster is vividly described in Nigel Calder's *Violent Universe* (1969), pp 64–5.
8. Dyson, 'Energy in the Universe'.
9. *Daily Telegraph*, November 11, 1969.
10. See, for example, R. Clarke, 'Technology for an Alternative Society', *New Scientist*, Jan. 11, 1973; M. Kenward, 'Alternative Technology—Politics and Yogurt?' Ibid.
11. R. Tsu, 'High Technology in China,' *Scientific American*, Dec. 1972.
12. Hermann Rauschning, *Hitler Speaks* (Thornton Butterworth, London, 1939), p. 220.

13. I must admit having borrowed this striking phrase from Bergier and Pauwels, *The Morning of the Magicians.*

APPENDIX III

1. F. J. Dyson, *Science,* 131, 1667 (1960).
2. F. J. Dyson, in *Perspectives in Modern Physics,* ed. R. E. Marshak (Interscience, 1966).
3. Ibid.
4. See, for example, J. M. A. Danby, *Fundamentals of Celestial Mechanics* (Macmillan, London, 1964).
5. Dyson, in *Perspectives in Modern Physics.*
6. Adrian Berry, private communication, 1972, after his discussions with Fremlin and Michaelis.
7. Dyson, in *Perspectives in Modern Physics.*
8. F. J. Dyson, 'Intelligent Life in the Universe', lecture given at San Francisco, September 18, 1972.

Index